SCIENTIFIC WINEMAKING
—made easy

Preface

THIS book has been inspired by the many friends I have made amongst the amateur winemakers of the United Kingdom. A great deal of effort is put into the making of home-made wines often with only a slight understanding of the principles underlying the practical work involved.

In this book I have attempted to explain in simple terms some of the principles scientifically and chart a path for the amateur winemaker who wishes to improve his or her efforts.

J.R.M.

Dedication

To friends who toil with the odds but produce the Goods.

Copyright is reserved

Acknowledgments

MY thanks are given to all those people who assisted in the preparation of this book. Their criticism, help and encouragement was much appreciated.

Special thanks must go to "Tilly" Timbrell who did most of the manuscript typing. In addition she rendered great help in reading and criticising each chapter so that in the final form it would be meaningful to the amateur winemaker. She was also largely responsible for compiling chapter eleven from numerous references, personal experience and my notes.

J.R.M.

Scientific
Winemaking
Made Easy

by

J. R. MITCHELL, L.R.I.C., F.I.F.S.T.

© "The Amateur Winemaker" Publications Ltd.
Andover, Hants.

ISBN 900841 42 7

First published	.	.	1969
2nd impression	.	.	Jan. 1970
3rd ,,	.	.	Sept. 1970
4th ,,	.	.	Nov. 1970
5th ,,	.	.	June 1971
6th ,,	.	.	Nov. 1971
7th ,,	.	.	Aug. 1972
8th ,,	.	.	Sept. 1974
9th ,,	.	.	Dec. 1975
10th ,,	.	.	Sept. 1976
11th ,,	.	.	Sept. 1977

Printed in Great Britain by:
Standard Press (Andover) Ltd., South Street, Andover, Hants.
Telephone 2413

Contents

9. BOTTLING:
   Bottles, corks, avoiding oxidation, dressing the bottle.

10. STORING AND SERVING:
    Storage, positioning, binning, decanting, serving, choice of glass, wines and food.

11. HOW TO EVOLVE RECIPES:
    Types of wine, table of ingredients 40 first-class modern recipes, making a simple balance.

12. TASTING AND GLOSSARY OF TERMS:
    How to taste, record cards, wine terms.

13. MICRO-ORGANISMS OF FERMENTATION AND SPOILAGE:
    Yeasts, moulds, occurrence and control, types of yeast, bacteria, acetobacter, malo-lactic fermentation, detection and prevention of infection.

14. TROUBLES AND CURES:
    Oily film, silky sheen, ropiness, hazes, surface growths, acid taste, astringent taste, darkening, metallic taint, crystals, overcoloured.

15. SIMPLIFIED WINE CHEMISTRY:
    Organic constituents, composition of wine, alcohols, glycerol, aldehydes, acids, carbon dioxide, esters, ketones, amino-acids, proteins, enzymes, carbohydrates, pectins, gums, tannins, colouring, flavour, metals, salts, sulphur compounds, pH, buffers.

16. USEFUL TEST PROCEDURES:
    Determining specific gravity, determining alcoholic strength, calculating proof, determining total acidity pH, pectin, microbiological haze, free sulphur dioxide, iron and residual sugar.

Introduction

"WINEMAKING"—mention this word in any company and one is sure of an interested response !

How many remember an old Aunt or Granny in the country, memories rosily tinged with nostalgic recollections of strong, sweet beverages and of visits transformed into "occasions" by the hospitable habit of a glass of home-made wine ?

Now the art of home-winemaking and the drinking of commercial wines has increased so tremendously since the war, the splendid necessity to stimulate and cleanse the palate with wine during a meal and to offer a glass of wine when friends call is no longer a luxury but an accepted part of living to so many families. Palates have become educated, certain standards recognised and the need to strive to improve endeavours has become imperative.

For many it is true to say the fascinating hobby of making wine has opened up new vistas: how friendships warm and blossom with the opening of a bottle costing so little! Meals become banquets, conversations uninhibited, and discussions stimulating. For introvert and extrovert alike—barriers are down!

What are the snags? One can easily spend a great deal of time and money on a hobby and not get good results and this is a waste of cash and effort. No one wishes to produce off flavours, unbalanced or unsound wines. The situation depicted in the advertisement about "B.O.," where not even one's best friend will disclose the unwelcome facts, is bad enough, but to be deleted from all Hosts' or Hostesses' lists as someone producing acetic or unsound wines is an even more tragic and personal problem!

Tread the middle path of fun and technique and learn the right principles to be followed in the art of winemaking! The interesting, unseen world of yeasts, bacteria, and enzymes is waiting to be understood and the reward of accepting that challenge is the anticipation and satisfaction at a job well done, coupled with the delight of savouring a good, sound, balanced wine. But, as with driving a car, it is necessary to learn good habits from the word "go." I hope that this book will help.

Introduction

CHAPTER 1

The way it should start

Sterilisation

ALWAYS start on the right foot and learn good habits from the word "go." When I started work a Senior Chemist remarked to me: "Cleanliness is essential in all branches of our work." I have always remembered his words, which are certainly true of the making of fermented products. This then, is our starting point.

Sterilising equipment

In this book the word "sterilise" will occur over and over again. Why has so much emphasis been placed on this? The answer is that the alcoholic fermentation of a particular starting material is best done by a yeast selected with both that material and the desired finished product in mind. Other yeasts which will be present in the air will only detract from the best course chosen and it is therefore desirable to eliminate them. Bacteria also presents hazards, for instance by turning wine into vinegar (as explained in Chapters 5 and 13). Sterilisation of equipment removes these hazards and in doing so a great deal of hit or miss from the subsequent fermentation.

Types of sterilisation

- *a.* Heat
- *b.* Use of chemical agents
- *c.* Radiation

a. Heat sterilisation

Apparatus made of stainless steel, glass and certain heat-resistant plastics such as polypropylene or polycarbonate—but NOT polythene—may be sterilised by the use of heat in one form or another.

Wet heat, that is to say, steam as generated in a pressure cooker or autoclave, is preferable to dry heat, such as that produced in an oven, because micro-organism tolerance to it is less and shorter sterilisation times can be employed.

The size of some pieces of equipment, such as gallon jars, forbids pressure cooker sterilisation and the domestic oven has to be employed or, if this is impossible because the material is unsuitable for heat sterilisation or an oven of sufficient size is not available, the use of chemical agents must be resorted to. Regardless of which method of sterilisation is employed, it is essential that all equipment to be sterilised is scrupulously clean.

Sterilisation in a pressure cooker

The equipment should be assembled into units, where possible, before sterilisation. That is to say, if a racking syphon is to be sterilised, place the cork on to the glass tubes and connect the glass to the plastic or rubber tubing, so the assembly is as in fig. 1.

Fig. 1

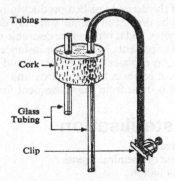

Tubing

Cork

Glass Tubing

Clip

12

The whole assembly can then be wrapped in Kraft (a heavy white paper) or greaseproof paper, secured with an elastic band and sterilised for a period of 20 minutes at 15 lbs. per square inch pressure (1.06 kg./cm^2).

Allow the cooker to cool slowly until the pressure is zero. The package of apparatus can then be removed but should not be opened until it is required for use. When the equipment is to be used, hands should be washed and care taken to avoid handling any surface which will come in contact with the wine or juice when assembling. P.V.C. polypropylene or polycarbonate plastic apparatus can be sterilised in this way —but NOT polythene.

Sterilisation in the oven

Remember to make sure that the equipment to be sterilised will withstand heat for a prolonged period of time. Old glass vessels will almost certainly break with the thermal shock. Plastic apparatus in its various forms will melt and produce an unwanted oven cleaning job!

If jars are to be sterilised, plug them with cotton wool. Pipettes or glass tubes can be wrapped in newspaper, taking care to screw up the paper wrapping at the ends so that when the articles are removed from the oven, they will be kept in a sterile condition.

Sterilise for one hour at 300°F. (150°C.). Bacterial spores can withstand higher temperatures of dry heat than wet and therefore the sterilisation time is longer and the temperature higher than for the pressure cooker method. Despite the reliability of modern oven thermostats, in my experience they do go wrong. Putting paper covered objects in an oven and then leaving them unattended is dangerous. Neither the fire brigade or the insurance agents will appreciate consequent calls upon their services, so stay in the house (after all, there are other things which can be done whilst watiing). Allow apparatus to cool slowly.

Sterilisation with hot or boiling water

This has distinct disadvantages over the other two methods just described but also has certain attractions. Water, to be really effective as a sterilant, must be at least 185°F. (85°C.) and boiling water 212°F. (100°C.) is to be preferred, since heat is lost from the water to the object being sterilised and the effective temperature falls more quickly than is realised. A contact time of 20 minutes has been found to give very good sterility using boiling water and small stainless steel apparatus. Swilling out with boiling water is nowhere near as certain to give sterility as completely filling the apparatus. With glassware the thermal shock of adding boiling water is much more severe than either the pressure cooker or oven method and a high risk of breakage exists. This method of sterilisation can be used for polypropylene, polycarbonate and P.V.C. apparatus.

A point often missed is when boiling water is tipped out of a vessel, unsterilised air immediately enters to fill it and if the vessel is not used at once it will need to be retreated when it is required. With the oven or pressure cooker method, if the equipment is cotton wool plugged and allowed to cool without opening the oven or cooker, the air contained is sterile and if necessary, the apparatus can be set aside and marked 'sterile' for use at a later date.

b. Chemical sterilisation

No thermal shock is involved with this method and the easiest and cheapest chemical sterilant for the amateur wine-maker to use is sodium metabisulphite. It dissolves more easily than tablets, has a higher sulphur dioxide content than potassium metabisulphite, and is cheaper to use.

To prepare a 1% solution with respect to sulphur dioxide —it is this which is the sterilising agent and is released when the salt is dissolved in water—add 2½ ozs. (or 74 gms.) sodium metabisulphate to 1 gallon of water.

Fill to the brim any apparatus which is to be sterilised and allow to stand for one hour; when the time is up, drain the apparatus by inverting on a previously cleaned draining board. DO NOT FLUSH OUT WITH TAP WATER for some supplies contain yeast and bacteria which, although harmless, can give wine spoilage problems. When the vessels have drained and are to be used, rinse them out with a little boiled and cooled water. This water will be "safe," the boiling having killed any micro-organisams.

Hypochlorite solutions, such as Milton and Chloros, are also good for apparatus sterilisation used as a 1 in 80 dilution but greater care must be exercised in rinsing them from the equipment or an unpleasant taste will arise in the products subsequently put into the apparatus. Milton used as suggested by the makers for sterilising a baby's feeding bottle is a good proprietary product viz, 1 tablespoonful of Milton to each quart of water.

c. Radiation sterilisation

This is an expanding field but because of the special safety requirements is of interest only to professionally employed winemakers and will not be dealt with here. Basically, the apparatus is exposed to radio-active rays which kill micro-organisms, and the equipment is thus sterilised. Ultraviolet light also kills micro-organisms.

Do's and Don'ts

DO USE a pressure cooker for sterilisation when possible.

DO USE an oven if a pressure cooker cannot be used but make sure the temperature is right.

DO wrap or cotton plug apparatus which is to be heat sterilised—it will then retain its sterility.

DO USE sodium metabisulphite solution for sterilisation without heat.

DO NOT put plastics into an oven.

DO NOT put old glassware into an oven or pressure cooker.

DO NOT go out and leave paper-wrapped objects in the oven.

DO NOT forget boiling water thermally shocks glassware and can break it.

DO NOT shock chill heat sterilised apparatus.

DO NOT use unboiled *tap water* for rinsing chemically sterilised apparatus.

CHAPTER 2

Selecting the right equipment

MOST beginners to the art of winemaking will probably wish to restrict the amount they spend to as small a sum as possible. The same remarks will apply to many people on limited budgets and these people have been borne in mind whilst writing this chapter, although consideration has been given to higher cost equipment.

Fruit pressing equipment

1. **Soft Fruit (blackcurrants, blackberries, strawberries, etc.)**

   In many cases this need be no more than cloths, a ribbed wooden board like an old-fashioned "dolly," a glass or wooden rolling pin, and a polythene bowl.

2. **Dried Fruit**

   This requires maceration and soaking, and consequently a sharp knife, a cutting board and a wide-mouthed jar are the only requirements. Dried Fruit can also be prepared by pressure cooking and this is dealt with in Chapter 3.

3. **Hard Fruit (apples, pears, etc.) or Citrus Fruits (lemons, oranges, grapefruit)**

   A small wooden wine press costing only a few pounds is a sturdy piece of apparatus, and the electric juice extractor is another excellent piece of equipment, designed to be useful in many ways in the kitchen. These appliances do improve juice yield.

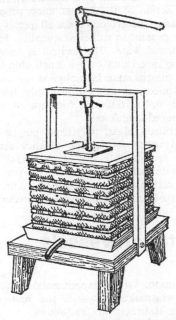

Fig 2. A small hydraulic, hand operated press

For the very affluent or for a winemaking club, larger hydraulic presses are available (fig. 2) but their cost is usually in excess of £40 unless one is lucky enough to acquire one second-hand. A word of caution: do not purchase equipment with copper or iron parts which are likely to come into contact with the juice. This remark applies from the simplest to the most complicated apparatus. Galvanised ware should on no account be used—it is dangerous, since zinc is dissolved by fruit acids to form soluble zinc salts which are poisonous. Copper and iron are likewise soluble in solutions of fruit acids and give rise to "off" flavours, bad colour, or deposits in the finished wine, so avoid them too. Copper contamination is the worst; copper salts pass through the whole fermentation process, virtually unaltered in quantity,

17

whereas iron contamination is considerably reduced by the natural fining action of the fermentation process. Ironware should nevertheless be avoided since all metals produce "off" flavours when present in sufficient quantity. For this reason do not use enamel ware. The enamel is easily chipped to expose the base metal and only a small chip starts trouble.

Coloured plastic material designed for use, say, in the garden or for toilet purposes will probably contain colouring matter which is not suitable for food use. Avoid, therefore, the use of coloured containers.

Clean, sterilised, glass, polythene, polypropylene, polycarbonate and other food-grade plastics are fine. Small wooden casks are suitable but need careful cleaning and this is dealt with below. Stainless steel, preferably in the polished state rather than the so-called "satin" finish, is of course, very satisfactory, being an inert material.

Fermentation vessels

Glass Vessels

In my opinion, for all experimental work on the scale of the home-winemaker, glass is the best material. My reasons for this statement are as follows:

1. The "activity" of the fermentation can be seen.
2. Glass is inert and cannot impart any "off" flavour or taint and the surface is completely non-absorbent.
3. Gallon jars are cheap and easily obtained. Carboys can also be purchased reasonably cheaply.
4. The level of cleanliness is easily seen and proper cleaning can be effected.
5. Racking can be easily controlled as the liquid/deposit interface can be seen at all times.

Because of their more robust nature large earthenware and wooden vessels become more attractive when volumes over five gallons are going to be processed as a single batch. Carboys should not be overlooked however. Lack of transparency with wood and earthenware means a need for greater care in cleaning.

Casks

Casks require careful cleaning. If they have been stood empty for any length of time they will also need to be reswelled to make the staves fit. The first step is to fill up such casks with water and allow the staves to tighten up. The hoops will then need to be hammered down towards the widest point of the cask right round its girth to give firmness. Scour out the cask with one-third of its volume of boiling water and put into it a clean two yard length of medium size link chain. Bung up and roll it about for 20 minutes or so. In addition to cleaning the cask the exercise is said to be good for the back and pectoral muscles, although many people have misgivings about this! Empty out the dirty water and the chain and rinse with clean water. To sterilise, fill the cask with 2% sulphur dioxide solution, made by dissolving 5 ozs. (142g) of sodium metabisulphite for each gallon of the cask's volume. Leave for 24 hours. This technique is more expensive than burning sulphur matches but is effective over the whole inside area of the cask. Sulphur matches not only produce sulphur dioxide but other products of combustion as well; these can lead to "off" flavours. The sulphur drips which form on the bottom of the cask beneath the burning sulphur match are troublesome, producing "off" flavours; a rotten egg aroma in a wine can sometimes be traced to this source of contamination when all other aspects of the processing have been conducted in a faultless manner.

Testing a Cask for Taint

After a cask has been cleaned and sweetened, as the operation just described is called, fill it with clean tap water, bung and leave for 24 hours, then taste the water to ensure no taint has been picked up.

An empty cask can be smelt for sweetness by removing the bung and smelling whilst striking the cask end a blow with the flat of the hand; this moves the air in the cask and any acetic or musty aroma is easily detected.

Storage vessels

Wooden casks need careful cleaning but when this has been done properly they are good containers, providing the wine has sufficient body to withstand the oxidation changes which can take place during storage in them.

Old beer or cider basks are fine for plants on the drive of a house but require re-coopering to make them suitable for wine storage. Do not use wax lined casks unless someone else has tried the same liner and found it satisfactory. The French wine industry has made several costly mistakes with wax linings and although a solution seems to have been found, for the amateur the experience could be costly in terms of wasted wine.

Polythene vessels for use as wine storage, in the opinion of the writer, fall into two categories.

1. **Wholly unsuitable**—because they smell strongly and will surely taint any wine placed in them. These can usually be easily detected by simply washing with water and then smelling. If any doubt exists, refill with water, stopper and leave for 24 hours. After the time lapse, take a glassful and taste it—if it tastes *exactly* like tap water the chances are the material will not taint the wine and the container comes in the second category.

2. **Suitable for wine storage for a limited period**—and four to six months is suggested as the maximum.

The reason for this is rather complicated but, putting it simply, the plastic because of its chemical composition and the heat forming process it will have undergone, will have the ability to remove from the wine on to itself some of the flavour constituents, thus robbing the wine of its "highlights."

Polythene is semi-permeable and this means molecules of sufficiently small size will pass through it. The writer has heard it said that alcohol will diffuse through polythene and drop the strength of the contained wine although he has no practical experience of this. What is certainly true is that air will diffuse through polythene and cause premature browning of even quite robust white wines, this tendency being much less marked with high density polythene than with the more common low density grade.

If a polythene container is to be used for short-term storage, do not forget to sterilise it with 1% sulphur dioxide solution allowing one hour's contact time. 2½ ozs. (71 gms) sodium metabisulphite powder per gallon of water will give a 1% sulphur dioxide solution. Glass, polypropylene, earthenware, stainless steel or wood provide excellent storage vessels, provided they have been properly cleaned and sterilised. All the reasons put forward on page 18 for selecting glass where practical, apply also to storage.

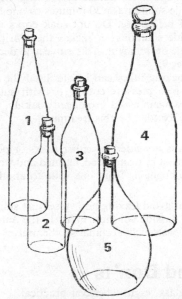

Fig 3. Bottle shapes in current use
1. German (Hock), 2. Bordeaux, 3. Burgundy, 4. Champagne, 5. German (Franconia)

Other equipment

Bottles

Try to use bottles which are suitable to the style and type of wine made, white and rosé wines in clear bottles, red wines in dark green bottles. If the aim has been a Bordeaux style, use a Bordeaux bottle, if a Burgundy, then a Burgundy bottle. The various bottle shapes are shown in fig. 3.

Always rinse out with 1% sulphur dioxide solution followed by boiled water and allow to drain. Bottling is dealt with in detail in chapter 9.

Corks, Plastic and Glass tubes for racking, etc.

Corks should be sterilised where possible in a pressure cooker or if this is not available by soaking the corks in 1% sulphur dioxide solution for 30 minutes to one hour, shaking the excess off before use. Do not soak corks indefinitely in sulphur dioxide solutions—it makes them brittle and more likely to break on removal. This subject is dealt with more fully in chapter 9.

Plastic racking tubes can be sterilised the same way as corks, either in a pressure cooker or with sulphur dioxide solution. Once again avoid prolonged soaking, as the plastic can become enbrittled by this treatment.

Taps

Where taps are fitted to either casks or polythene containers these must be dismantled, cleaned and sterilised before use and when changing from one wine to another.

General

In general avoid leaving any vessel just standing around with sulphur dioxide solution in it. If the solution is very dilute stray yeasts will acclimatise themselves and troubles could result.

Do's and Don'ts

DO use glass vessels whenever practical.

DO sterilise all equipment used for pressing, fermentation and storing.

DO use sodium metabisulphite solutions instead of sulphur matches.

DO avoid leaving vessels lying idle with dilute sulphur dioxide solution in them.

DO NOT use dirty or unsterile equipment.

DO NOT on any account use galvanised apparatus.

DO NOT use copper, brass, bronze, cast iron, mild steel or enamel appliances.

DO NOT use coloured plastics unsuitable for food use.

DO NOT forget to *clean* and sterilise taps fitted to vessels.

CHAPTER 3

Extracting and adjusting fruit juices for fermentation

IN this chapter juice derived from fruits, fresh, dried and in concentrated juice form will be dealt with.

Many amateur winemakers, in addition to the above juices, use flowers, leaves or vegetables to impart a flavour which can, under favourable conditions, give a pleasant finish to a fermented wine.

Essentially a sugar, acid and water solution is prepared and used to extract the flavour from the selected botanical material by simple soaking, or sometimes by leaving the flavouring material in the juice during the first stages of fermentation, thereby obtaining a better extraction with the alcohol produced.

Whilst not dealing with this type of wine at any length in this book the author does not wish to discourage those who make satisfactory wine by this method, but to those new to wine making it is only fair to say that wine produced in this way is rarely as good as that made from fruit juices or concentrates.

Juice sources

These can be summarised as follows:

Fresh fruit:

This group is sub-divided into three classes:

1. Soft fruit consisting of berries, bananas, etc.

2. Hard fruit such as apples and pears.
3. Citrus fruit, of which oranges, lemons and grapefruit are examples.

Dried fruit:

Such fruit as raisins, sultanas, figs, dates, elderberries, bilberries and apricots come into this category and have very high sugar contents, although apricots produce very "gummy pulps" on rehydration.

Concentrated juices:

These are obtained by evaporating a fruit juice down from its natural sugar strength to three, five, seven or even ten times the original concentration. Depending on the way in which the concentration is done, that is to say, with reduced pressure and recovery of volatile components, or crudely in open-pan evaporators, so will the quality of the concentrate vary. Concentrates can be used not only for diluting and fermenting but also for sweetening a wine after fermentation and fortification.

General comments on juice production:

The amount of fruit to make a wine will vary depending on if the wine is to be light or full bodied, table or dessert. A general method to prepare a juice suitable for fermentation is as follows:

The desired amount of fruit is crushed and added in a sterilised vessel to two pints of boiled and cooled water for each gallon of wine to be made and then enough sodium metabisulphite to cover a shilling is added for each final gallon, then after approximately four to six hours the mixture is well stirred, a sample taken, strained, and the specific gravity determined. The amount of sugar necessary to produce the desired alcohol content can then be obtained from Table 1. This is weighed out and added. When all the sugar has dissolved the volume can be made up to one gallon (or multiples), the total acidity determined and the amount of acid required calculated by the method described later in the chapter.

24

After thorough mixing the juice is then ready, except for the addition of pectic enzyme and yeast growth promoter, for fermentation.

Each section in this chapter describes the best approach for individual types of fruit but this is the basic method for all of them.

Obtaining the juice from fresh fruit

1. SOFT FRUIT

Many soft fruits are extremely difficult to press. Applying pressure to the fruit produces a "pulp" rather than a free running juice. For fruit of this nature, viz., strawberries, bananas, raspberries, blackberries, and loganberries, a good technique is to "pulp" the fruit, or if a food mixer is available, liquidise it and add 1½ volumes of water for each volume of fruit together with 1 gm. of sodium metabisulphite powder for each gallon of combined "pulp" and water. A simple way to do this and avoid mistakes is make a calibrated jar, or jars, into which the pulped fruit can be put as it is being made.

At a convenient point one can then stop, see how much is in the jar and add the requisite amount of water. A wide mouth—say, 2" to 3"—on the jar selected makes additions easy. Any jar can be calibrated and the finished result is shown in fig. 4. The following method will enable the reader to see how this simple job is done.

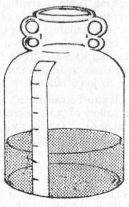

Fig 4

Take the jar you intend using and stick a 1" strip of white paper down the side from the bottom to the neck. Add ½ pint of water and mark its level on the paper

25

strip with a pencil or waterproof ink. Continue to add ½ pint quantities and mark off the levels until the jar is full. Waterproof the strip by either applying liquid wax or a strip of wide transparent adhesive tape. Wash the jar out and sterilise with 1% sulphur dioxide solution prepared as described in chapter 1. If fruit is being used in the crushed form, place it and the water into the calibrated jar. Note the volume of juice.

Add 1 gm. of sodium metabisulphite for each gallon of juice, add enough pectic enzyme to ensure no haze problems later (see the directions on the bottle), stopper and set the jar aside for four to six hours to allow the sugar of the fruit to be extracted by the water, then remove a sample and determine the specific gravity using a glass hydrometer, as described in chapter 16. The sample taken may need to be strained through a sieve or screen to remove the larger particles of fruit before the test is carried out. You now have the specific gravity and volume of the juice you intend to ferment. It is next necessary to decide the style of wine to be made. If a dessert wine is required, a higher starting specific gravity will be necessary to give a greater alcohol content in the finished wine than for a table wine. The relationship between starting specific gravity and potential alcohol content is given in the following Table 1. The figures are based on experimental experience and calculation of sugar content is accurate to ¼ oz: also in calculating the potential alcohol content allowance has been made for dissolved solids other than sugar by deducting 30 gms./litre (4.82 ozs./gallon) from the theoretical sugar content. In practice the dissolved solids will vary from juice to juice but this is a good working figure.

It can be calculated that each pound of sugar added to one gallon of juice increases the volume by approximately ½ pint or 6¼%. Obviously a gallon of juice cannot be taken in a gallon jar and sugar added to it because the liquid would overflow and the alternative procedure of taking a weight of sugar and making it *up to* one gallon is a much more convenient method. Column 4 of the table shows the weight of sugar *in* one gallon of juice corresponding to the specific gravity in column 1 after allowing for other dissolved solids as already described.

TABLE 1

Column 1	2		3		4		5
Specific Gravity @ 60°F (15.5°C)	Weight of Sugar Required per		Volume after Sugar addition to:		Weight of Sugar in		Potential Alcohol % by Vol.
	1 Gallon	1 Litre	1 Gallon	1 Litre	1 Gallon	1 Litre	
	lbs. ozs.	gms.	gall fl ozs.	Litres	lbs. ozs.	gms.	
1.020	— 8½	52	1 6½	1.041	— 8½	51	1.0
1.030	— 13½	83	1 9½	1.059	— 12½	78	2.6
1.040	1 2	112	1 12½	1.080	1 0¾	104	4.3
1.050	1 6½	143	1 16	1.100	1 5	131	5.9
1.060	1 11½	173	1 19½	1.120	1 9	156	7.6
1.070	2 0½	204	1 22½	1.141	1 13	181	9.2
1.080	2 6	237	1 25½	1.159	2 0¾	204	10.9
1.090	2 11½	271	1 29	1.181	2 5	231	12.6
1.100	3 1½	309	1 32	1.200	2 9¾	260	14.3
1.110	3 7½	346	1 35	1.219	2 14	287	15.9
1.120	3 14	387	1 38½	1.241	3 2½	315	17.6

N.B.—Each pint contains 20 fluid ozs. and 1 gallon equals 4.546 litres.

Adjusting the Sugar Content

From the specific gravity of the juice it is possible to obtain the potential alcohol content by reference to Table 1. (columns 1 and 5 respectively). Now what the amateur winemaker surely finds hard to accept is that grapes and only grapes, contain sufficient sugar to give a reasonable alcoholic content in the finished wine—even they need help on occasion; the addition of sugar to improve the alcohol content of wine (Chaptalisation) is practised in France and is confirmation that grapes are not always high in sugar. Having seen and accepted that sugar addition is necessary, a scientific correction of the juice can be made. The first thing to do is to decide what alcohol content is desired in the finished wine, remembering that 15% by volume is just about the maximum alcohol content the amateur can expect to achieve, although this is not the maximum possible, as will be explained in a later chapter. Thus a starting specific gravity over 1.100

will almost certainly result in sugar remaining after fermentation has ceased, which is only of use if a sweet dessert wine is desired; thus a lower starting specific gravity should be selected for a semi-sweet or dry wine. The alcohol content in the finished wine decided, the amount of sugar to be added can easily be calculated as follows:

Using the figures in column 4 of the table subtract from the sugar content corresponding to the % potential alcohol required (or if a dessert wine is desired with residual sweetness, the sugar content corresponding to an S.G. of either 1.110 or 1.120) the sugar content already present in the juice. The answer is the amount of sugar required to make one gallon.

For example:

Say a juice had a specific gravity of 1.030, referring to the table (column 4) this corresponds to a sugar content of 12½ ozs. (354 gms.) per gallon.

If a table wine with an alcohol content of 12.5% approx. is required, this corresponds to 2 lbs. 4¾ ozs. (1,040 gms.) of sugar per gallon.

2lbs. 4¾ ozs. (1,040 gms.) − 12½ ozs. (354 gms.)
$$= 1 \text{ lb. } 8¼ \text{ ozs. } (686 \text{ gms.})$$

Thus, 1 lb. 8½ ozs. (686 gms.) of sugar is required for each gallon of juice and this can be weighed out, placed in the fermentation jar, and the juice added to make the volume up to a gallon (or multiples thereof, as the case may be).

The Role of Acidity

Acidity is essential to give a wine good keeping properties and resistance to bacterial attack, factors which are fully discussed in later chapters. All fruit juices contain acids to a greater or lesser extent but with the exception of grapes and some other soft berry and citrus fruits, the acidity is rarely high enough to give sufficient protection or acidity for winemaking. It is therefore necessary to add acid for this purpose and citric acid is the acid generally chosen, although malic acid has found favour amongst some winemakers.

Acidity Correction

Matching acidity to the desired finished style of the wine is, of course, important and to assist in doing this Table 2 has been prepared. It shows the acidity of various styles of wine calculated in terms of % citric acid. Those who wish to use malic acid should note that it is organoleptically 20% more acid than citric but the same figures can be used to calculate the necessary additions; then deduct $\frac{1}{5}$ of the total weight.

TABLE 2

Wine style	Titratable acidity as % citric acid
Dry red	0.57
Dry white	0.70
Medium dry Rosé	0.75
Medium dry white or red	0.65
Sweet white table	0.70
Sparkling	0.80
Sherry type, sweet or dry or dessert white	0.50
Port type or red dessert	0.45

Having decided the style of wine which you wish to make, measure the total acidity of the juice *after* the sugar addition. The test is explained in chapter 16.

From experimental evidence 1 lb. of B.P. citric acid raises the acidity of 10 gallons of juice by 1% (or 1.6 ozs. to 1 gallon). Thus, if the total acidity of the juice is known and the desired finishing acidity is selected from Table 2, it is an easy matter to calculate how many ounces of citric acid per gallon of juice to add, as follows:

To Calculate the necessary addition of Citric Acid

Subtract the total acidity of the juice from the total acidity desired in the finished wine. Multiply the answer by 1.6 and this is the number of ounces of citric acid B.P. required for each gallon of juice.

In the case of gram additions to a litre simply multiply by 10.

viz:

(Total acidity desired − Total acidity of Juice) × 1.6 = ozs/gall.

alternatively

(Total acidity desired − Total acidity of Juice) × 10 = gms/litre

Example:

A dry white wine is to be made and the juice has a total acidity of 0.35% citric acid. The acid addition required equals (0.70 — 0.35) × 1.6 ozs. or 0.35 × 1.6 ozs. citric acid per gallon = 0.56 ozs. (Just a little over a half ounce) per gallon.

The most satisfactory method of adding the citric acid is to extract a volume of juice, dissolve the acid in it and then pour this juice back into the bulk and mix well.

It is possible to use acids other than citric or malic for acidity corrections but for the amateur there are usually objections. It has to be stressed, however, that the juice *must* be sulphited to prevent bacterial attack on the citric acid.

Tartaric Acid

Tartaric acid is the most predominant acid in grape juice but its addition to acid deficient juices can give rise to deposits of bitartrate later in the life of the wine, especially in hard water areas or where potassium metabisulphite and not sodium metabisulphite has been used as a source of sulphur dioxide. The reason for this is that tartaric acid slowly combines with any potassium or calcium present in the juice to form either cream of tartar (potassium hydrogen tartrate) or calcium tartrate. The latter is the most insoluble of the two but both will give deposits of crystals if left sufficiently long. A sudden cold snap will hasten crystal formation in wines stored without any thermal protection.

Malic Acid

Apart from having a different flavour, this acid requires careful use since it can in part be changed by a bacterial fermentation to lactic acid which has a characteristic sour, smooth flavour which although desirable in small quantities is undesirable in other than traces. The process of malolactic fermentation is explained in chapter 13. The presence of "free" sulphur dioxide will usually prevent this malolactic fermentation from taking place. This is the reason why in great red wines the sulphur dioxide content is kept low thus allowing the change of malic acid, which is rather tart, to lactic acid, a smoother and less acidic acid, thereby giving a smoother rounded effect to the wine. Malic acid can be used instead of citric acid but sufficient sulphur dioxide *must* be available to prevent the malolactic fermentation taking place at an inopportune time—this is not easy for the amateur to guarantee.

2. HARD FRUIT

Sufficient juice is present in this type of fruit for fermentation without the addition of water. A press or juice extractor is necessary to obtain an economic yield of juice and the fruit must be cut up to form a "pulp" before pressing and, if this is put into a fine weave muslin or terylene cloth, losses will be reduced and disposal of the pomace (fruit after juice extraction) made easy. When an electric juice extractor is used the fruit will reduce to a purée rather than transparent juice. However, if it is sulphited to prevent bacterial attack and oxidation, as indicated below, then left to stand covered with a saucer or sterilised cloth for 12–24 hours it will be found that the solid has separated to the bottom and the now nearly clear juice can be decanted off. If necessary the juice can be extended by the addition of one volume of water to every two volumes of fruit juice.

The specific gravity and total acidity of the fruit juice or juice plus water, as the case may be, should then be tested, the sugar and acid additions worked out as described in the last section dealing with soft fruits, and the additions made and mixed in. It must be remembered that the acidity correction should be calculated on the basis of the volume *after* sugar addition because the sugar addition in effect decreases the acidity of the juice (or juice and water). One gm. of sodium metabisulphite should be added for each gallon of sweetened juice to prevent bacterial action and oxidation.

3. CITRUS FRUITS

For this group of fruit a different approach is needed. The skin of this class of fruit is particularly rich in essential oils (see chapter 11) and these are toxic to yeasts. That this is so can easily be demonstrated by putting some chopped orange peel into a vigorous fermentation; within a short time it will have slowed right down. This is important in deciding the best method of expressing the juice of citrus fruits.

Whole ripe fruit only should be used, as over-ripe fruit lowers the quality of the final wine, giving a jaded effect. Washed whole fruit cut in halves or quarters and either squeezed by hand or in a press will give a good yield of juice.

Correction of Acidity and Water Addition

Citrus fruit juice is high in acid and to obtain the correct acidity for winemaking needs to be extended by the addition of water. Dry white wines of a clean elegant nature can be made from citrus fruit and a suitable starting acidity can be arrived at as follows:

1. Test the juice for total acidity as described in chapter 16.
2. Decide the most suitable acidity from table 2.

A quick calculation to decide the amount of water required can be done as follows:

Write down the total acidity of the juice on the left-hand side of a piece of paper and the acidity required (from Table 2) beneath it. Take one away from the other. On the right-hand side of the paper write the acidity required again. Omit the decimal points and on the *left* will be the volumes of *water* to correct the acidity of the juice volume on the *right*.

Example:

A grapefruit juice had an acidity of 0.92% as citric acid, and a dry white wine with an acidity of 0.70% as citric acid is required.

Water	Juice
0.92	
0.70	
0.22	0.70

Leave out the decimal points

22	70

For every 70 volumes of juice, 22 volumes of water need to be added to obtain a juice with an acidity of 0.70% citric acid. From this, fractional parts are easily worked out, i.e., 7 volumes juice to 2¼ volumes (approx.) water.

Using this method the acidity is easily corrected.

In the case of citrus fruit the specific gravity of the juice has to be measured *after* the addition of water to correct the acidity. The percentage alcohol desired in the finished wine is decided upon and the corresponding sugar content *in* a gallon found from column 4 of table 1. After determining the specific gravity of the juice plus water, its sugar content is looked up in the same fashion and the difference between the two values is the sugar required to prepare a gallon of juice. The calculation is simple and the same as explained in the hard fruit section, viz.:

Sugar Required for 1 Gallon of juice = Sugar content for % alcohol required – Sugar content present in the juice plus water.

The sugar should be placed in the sterilised fermentation vessel and the juice added with mixing until the volume is approximately one gallon. Citrus fruit juices provide particularly good media for the growth of micro-organisms, having many of the nutrients necessary for rapid yeast multiplication. Also, unless protected by the addition of sulphur dioxide, rapid oxidation sets in and the juice becomes rather stale before fermentation is completed. For these reasons 1½ gms. of sodium metabisulphite should be added to each gallon of juice *after* the addition of water and sugar.

Obtaining the juice from dried fruits

The technique for dealing with dried fruit is different from that used for any of the other types of fruit so far discussed. The first essential is to leach as much sugar and flavouring from the fruit as possible. However, in view of the fact that dried fruit will have been stored under a variety of conditions, between the time it was fresh to the time it reaches the consumer, it is likely to have a large micro-organism population which, if given a favourable opportunity, will flourish and spoil any winemaking efforts. This problem can be overcome in two ways:

1. Steeping the fruit in boiling water

or

2. Pressure cooking the fruit and water together.

The Boiling Water Technique

Every one pound of dried fruit will require four pints of boiling water to rehydrate the fruit and extract the sugar. Whilst the water is being heated the fruit should be washed and where possible (bearing in mind elderberries, bilberries, etc.) shredded into small pieces and placed in a sterilised fermentation vessel. The boiling water should then be added and the jar set aside, loose stoppered with a cotton wool plug moistened with 2% sodium metabisulphite solution, to cool. When cold, 1½ gms. of sodium metabisulphite should be added for each gallon of juice, the acidity measured and citric acid added to give the desired acidity. The specific gravity is then measured. Because of the low concentration of flavouring substances present in dried fruit, it is common practice to ferment the juice together with the fruit at this stage and, in so doing, extract, with the alcohol formed during fermentation, a greater amount of flavouring substances. As soon as the fermentation begins to slow, the juice is tested for specific gravity and the alcohol content worked out, the juice strained off from the fruit residues and sufficient sugar added to give the desired alcohol content in the finished wine.

Table 1 can be used to calculate the sugar addition necessary when the alcohol content has been determined (See chapter 16 for the method of determining alcohol).

The Pressure Cooker Technique

The only difference between this method and the one above is that the prepared fruit and water is placed in a pressure cooker which is then raised to 15 lbs./sq. in. pressure and maintained for 10 minutes. When the cooking time is over, the cooker can be chilled under a stream of cold water and the juice treated exactly as in the boiling water method.

Dried peaches and apricots can present problems as the water extract tends to contain a high concentration of gums.

Obtaining the juice from concentrates

The degree of concentration varies considerably from concentrate to concentrate and the most satisfactory way of determining the correct dilution is to use the acidity of the material as the basis from which to work.

The procedure is first to determine the total acidity of the *concentrate* and then decide with the help of Table 2 what acidity is required in the finished wine. Using the same calculation as explained for the correction of acidity in citrus fruit juice the proportion of water to concentrate can be determined.

Example:

If a concentrate had a total acidity of 3.6% as citric acid and a wine with a finished acidity of 0.70% as citric acid was desired, proceed as explained on page 32, viz:

3.6 Concentrate
0.7 Required
———
2.9 Difference   0.7 Required

Omitting the decimal points.

29 volumes of water to 7 volumes of concentrate will give a juice with a total acidity of 0.7% as citric acid.

Or roughly 4 of water to 1 of concentrate.

After the addition of water for correction of total acidity, the sugar content of the juice will require correction. This is easily achieved by determining the specific gravity of the juice plus water and finding the corresponding sugar content from table 1, column 4 as already explained.

After looking up the sugar content for the alcohol required the calculation previously explained is carried out, viz:

Sugar required for 1 gallon juice = Sugar content for % alcohol required — Sugar content in concentrate plus water.

The sugar is put into the fermentation vessel and juice added with mixing to make one gallon.

One-and-a-half gms. of sodium metabisulphite/gallon of juice should be added to prevent bacterial infection.

Useful fermentation additives

The remarks made in this section apply to all of the various juices and juice sources discussed earlier in the chapter.

Depectinising Enzymes

Pectin is a substance which occurs to a greater or lesser extent in the juice of all fruits. It is responsible for the gelling of jams and if present in a finished wine will give rise to a hazy appearance. Chemically pectin consists of complex units of a-galacturonic acid and its methyl ester.

```
        CHO                              CHO
         |                                |
   H— C —OH                        H— C —OH
         |                                |
  HO— C —H                        HO— C —H
         |                                |
  HO— C —H                        HO— |  —H
         |                                |
   H— C —OH                        H— C —OH
         |     1 unit of Pectin        |     Methyl ester of
      COOH  ∝-Galacturonic         COOCH₃ ∝-galac-
                    acid                        turonic acid
```

This acid in its basic form has many similarities in chemical structure to sugar which, of course, ferments to form alcohol and other products during the production of a wine. Similarly much of the pectic substances present in fruit juices are broken down during the course of fermentation.

However, some juices, in particular those of citrus fruits, have a great deal of pectin all of which is not destroyed by fermentation. This is when a pectic enzyme is particularly useful as it will promote the breakdown of the complex pectin structures to more simple ones which are either completely soluble and present no problem or are sufficiently insoluble to precipitate and can be removed by racking. Generally, if any doubt exists, it is best to add a pectic enzyme to the juice before fermentation rather than omit it. Many companies market enzymes and selection is largely a matter of experience. Generally liquid enzymes do not retain their activity when stored for any length of time.

Solid enzymes which have to be used in relatively large quantities, have usually been absorbed into a carrier agent and this has the effect of diluting the enzyme and, apart from it being undesirable to add unnecessary substances to a fermentation, the cost is higher for a lesser amount of actual enzyme material.

The worst pectic enzymes are those packaged badly, since they can carry micro-organisms most undesirable to a fermentation. The best guarantee is to buy the most concentrated enzyme of a reputable company and store for as little time as possible, thus buying fresh material rather than relying on old stock.

Yeast Food and Promoters

It has been known for a long time that yeast activity can, under some circumstances, be increased by the addition of chemical materials; broadly speaking these are as follows:

Ammonium salts

Ammonium salts contain nitrogen and this is used by yeast cells to multiply. Generally a source of nitrogen is the only thing which need be added to a fruit juice and ammonium sulphate B.P. $(NH_4)_2SO_4$ provides a cheap and easy source. An addition of three grammes to each gallon of juice will adequately supply the yeast's nitrogen requirements not fulfilled by the juice.

Phosphates

For a fermentation to proceed, some phosphate material is essential. Normally a sufficiency exists in a fruit juice or extract but sometimes a fermentation will halt (or stick)

37

when all other conditions such as temperature and aeration have been considered. When this happens the addition of about as much ammonium phosphate B.P. as will cover a new penny (or $\frac{1}{2}$ gm.) followed by a vigorous shaking, will often in a short while get the fermentation going again.

The reason for phosphate requirement? When sugar starts to ferment, an ester containing phosphate is formed which, as will be seen later, forms an essential step to the process of fermentation.

Ammonium phosphate $(NH_4)_3PO_4$ also contains nitrogen as part of its chemical structure. This also is utilised by the yeast just as in the case of ammonium sulphate hence ammonium phosphate can if desired be added instead of ammonium sulphate.

Vitamin additions

Some commercial brewers use mixtures of vitamins and ammonium salts but the practice has not found wide favour with winemakers, adequate amounts usually being present in the juice. However, Thiamine (Vitamin B_1) will increase the activity of yeasts especially in stuck fermentations and it has been suggested that the transformation of alcohol to acetic acid esters which give rise to volatile acidity is slowed down by the presence of thiamine. For those who wish to use Vitamin B_1 as a fermentation additive, the amount necessary to cover the *tip* of a penknife blade is sufficient for a gallon of juice. Of the other vitamins little conclusive evidence exists as to their usefulness in promoting yeast activity or producing any other desirable side effects during fermentation.

Summary

The addition of ammonium sulphate B.P. as a nitrogen source and thiamine as a growth additive is generally sufficient for any amateur. Ammonium phosphate may prove useful as well but little purpose is served by other additions.

Do's and Don'ts

DO remember that different fruits and juices require different preparation.

DO use depectinising enzyme, it gives assurance.

DO add ammonium phosphate or ammonium sulphate and possibly thiamine.

DO NOT leave juice without an addition of sodium metabisulphite; it prevents bacterial infections.

DO NOT be put off by a bit of theory—it helps to understand the practice.

DO NOT omit the need to test for specific gravity and total acidity.

DO NOT forget to correct juices for sugar and acid content.

DO NOT be discouraged.

CHAPTER 4

Preparing the yeast starter

MOST of the principal wine producing areas of Europe rely upon the naturally occurring indigenous yeasts for fermentation and without any doubt the success achieved is a test mony to the system. However, the South African, Australian and Californian wine industries use cultures of wine yeasts produced under laboratory conditions to start fermentations and the commercial success of their enterprise should not be overlooked or passed over lightly. Amateur winemakers in the United Kingdom do not have the advantage of indigenous wine yeasts and the population of other microorganisms is such as to make a starter of cultured yeast essential if good results are to be achieved.

Yeast culture forms available

1. Baker's yeast—this is a stiff paste stored by refrigeration.
2. Dried yeast—sold in packet form or compressed into tablets.
3. Liquid yeast cultures in saline, sugar or other preservative solutions.
4. Yeast streaks on agar or gelatine culture media.

Baker's yeast, whilst carrying out a fermentation, does so without any finesse and although satisfactory for the beginner the "bready" aroma and flavour which tend to be imparted to the wine are undesirable for any finesse.

40

Dried yeast requires great care in the drying and packaging to ensure the yeast strain remains true. Unfortunately, contamination often creeps in. Some skill is involved in developing cultures from dried yeast but they give quite good fermentation activity. "Off" flavours are sometimes developed with dried yeast and they do not sediment as well as fresh yeast.

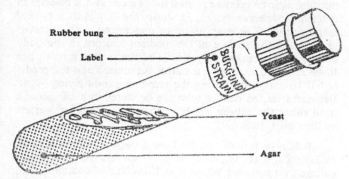

Fig. 5

Liquid cultures are usually exactly what they are claimed to be by the vendors, as are cultures on *test tube slopes of nutrient agar or gelatine*. These two forms of culture are to be preferred to dried yeasts: a certain amount of technique is called for in using them and this will be explained later. The cultures in either of these two forms can easily be stored in a refrigerator and will last for a long time because of the sterile method of packaging.

Preparation of the starter

Dried or baker's yeast starter

For those who prefer to use dried yeasts or baker's yeast, the best approach is to take one teaspoonful of yeast for each gallon of juice to be fermented. Rehydrate dried yeasts or make baker's yeast into a slurry by using some of the juice which is to be fermented and then mix this into the main body of the juice.

A starter from an agar or gelatine slope

This type of culture is really the cheapest because only very little is used at a time and it can be stored in a refrigerator for about six to nine months before the media dries.

Compared with a dried yeast or even a liquid culture, a nutrient agar or gelatine culture looks more as if it belongs in a scientific laboratory. Fig. 5 shows the details of a typical slope culture. The main thing to understand is the yeast is growing on the surface of the medium and not in the body of it. For this reason only a little of the surface growth has to be removed each time a starter is required and the media is left to carry on feeding the remaining multiplying yeast. Between uses the culture tube can be stored in a refrigerator until either the culture has been completely used for starters or the media dries out.

It is now possible to see how a slope culture is used in making a starter. The working up of an agar slope or liquid culture into a starter which is still true and uncontaminated, calls for the use of a clean, sterile apparatus and careful working. Some simple equipment is necessary so start by getting this together.

First of all, an innoculating needle is required. The simplest form is shown in fig. 6. It consists of a stiff single strand of wire of 26 S.W.G., or thereabouts, which is mounted into or bound on to a handle, which can be wooden or metal. Special holders can be purchased from a laboratory supplier for about 40p. The wire can be found, if no other source exists, in an old electric light bulb. The two support wires which carry the actual filament are ideal.

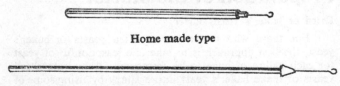

Home made type

Laboratory type
Fig. 6

An old pen handle will do for mounting the wire. Hold the wire in a pair of pliers, heat 6 mm. of the wire to red heat in a flame and drive it into the wood. A loop is then made in the other end of the wire.

Secondly, two wide mouthed bottles—a 150 ml. and a 300 ml.—or two conical laboratory flasks of 150 ml. and 300 ml. capacity are necessary. (See fig. 7) together with some rubber bungs.

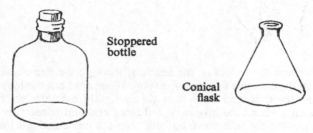

Stoppered
bottle

Conical
flask

Fig. 7

All that is needed in addition, is a spirit lamp, primus or picnic stove, gas jet or Bunsen burner.

The procedure is then as follows:

Into both of the wide mouthed containers put 100 mls. of juice, prepared as in chapter 3 for the type of fruit and style of wine to be made but *without* any sulphur dioxide, which at the early stage the yeast will be too sensitive to withstand. To the juice in the 300 ml. bottle, add enough sodium metabisulphite to cover the *tip* of a penknife, and to both bottles enough ammonium phosphate B.P. to also cover the *tip* of a penknife. Cotton wool plug the bottle mouths (do not screw stopper) and sterilise the bottles and contents in a pressure cooker at 15 lbs. per sq. in. for 15 minutes.

Allow the sterilised juice to cool to room temperature. Heat the wire loop in a flame to red heat, allow it to cool but not touch the bench or table you are working on. Take the culture tube in your left hand and remove the rubber bung or cap with your little finger as shown in Fig. 8.

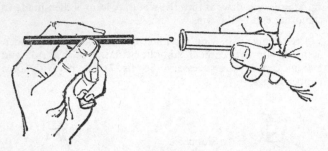

Fig. 8

Pass the mouth of the test tube through the flame and then do the same with the wire loop. When cool put the loop into the tube and draw it relatively lightly over the yeast culture. Flame the tube neck and bung end and reinsert the bung. Put the tube down but still keep the innoculating wire in your right hand. Remove the bung of the 150 ml. bottle or flask in the same way as before, flash the mouth through the flame and plunge the wire loop—not the complete wire —into the juice and move it backwards and forwards a couple of times to remove the yeast. Flash the bottle mouth and bung through the flame again and reinsert bung.

Give the bottle a thorough shake and set aside in a warm place (70–80°F.) for 24–48 hours—keep an eye on it and when it is fermenting nicely, remove the plug from the bottle, put it down and flame the neck, remove the plug from the 300 ml. bottle, flame its neck and after swirling, tip the contents of the 150 ml. into the 300 ml. bottle. Reflame and plug. Shake vigorously and store at 70–80°F. and when fermentation activity is strong use what is now a strong yeast culture to start the main fermentation.

From a liquid culture

The fundamental difference between an agar slant and a liquid culture is that whereas yeast cells are growing on the surface of nutrient agar, the rate of growth depending on the nutrients present and the temperature of storage, liquid cultures, usually in 10% sucrose (cane sugar) are in the resting stage. Working up a liquid culture requires similar care to that described for an agar slant if contamination is to be excluded.

Prepare the sterile juice in 150 and 300 ml. bottles exactly as described above. Uncap the yeast culture and the 150 ml. bottle and after quickly flaming the bottle mouths, tip the yeast culture into the sterile juice. Reflame the bottle and rubber bung quickly and reseal the bottle. Shake vigorously for about a minute to aerate the juice well—this is important for the yeast requires dissolved air to multiply, as will be explained in the next chapter. Set the culture aside in a warm spot at between 70–80°F. and when fermentation activity is strong, transfer the whole bulk to the 300 ml. bottle, taking care to do the operation under as sterile conditions as possible, viz. flaming the necks of the bottles and the bungs and working as quickly as possible. Shake the bottle and allow the culture to increase at 70–80°F. as before. When activity is strong, the starter is ready for use

Using either method, the first stage of fermentation has then been completed.

The main fermentation will require 4–6% of its volume of starter. Ten fluid ounces is sufficient for one gallon (being just over 6% of the total volume).

Do's and Don'ts

DO try to use yeast cultures—they do give good results when used properly.

DO give attention to sterilisation details.

DO make sure to aerate yeast starters by shaking.

DO NOT regard the starter as unimportant—for quality results it is most important.

DO NOT be afraid to use a yeast culture—attention to detail is all that is required.

DO NOT disregard the need for "sterile" working—this is the secret of good results.

DO NOT regard baker's yeast as the only one for you.

CHAPTER 5

Managing the fermentation

IN this chapter, after starting off the practical aspect of the fermentation that has been prepared via the sterilisation of equipment, balancing the juice for sweetness and acidity and the propagation of the starter, the simplified theory of fermentation and yeast activity is discussed.

Getting the fermentation going

Before doing anything else, make sure the fermentation vessel, bung and trap have been sterilised by one of the methods described in chapter 1. Fig. 9 shows alternative traps which can be used and how they operate. With very vigorous fermentations a loose cotton wool plug moistened with 1% sodium metabisulphite solution is quite adequate but when the fermentation slows it is advisable to replace this with one of the types shown opposite.

A. is made of plastic and robust.

B. is constructed of glass and, whilst very effective, is fragile.

C. is home-made. It consists of a glass tube which goes through the bung to which is attached a piece of rubber or plastic tubing. The end of the tube dips about 12.5 cms. below the liquid held in a small bottle, tube or jar, which is strapped to the neck of the fermentation vessel by adhesive tape. Although less elegant, it is cheap and effective.

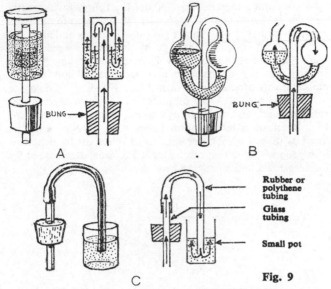

Fig. 9

A

B

Rubber or
polythene
tubing

Glass
tubing

Small pot

C

In either pattern a 1 % solution of sulphur dioxide should be used for the trap liquid rather than water. It should, however, be replaced by a fresh solution if more than one month is taken by the fermentation because the carbon dioxide gas passing will by then have lowered the sulphur dioxide concentration. Micro-organisms are much more likely to flourish, even towards the end of the fermentation, if the sulphur dioxide concentration has considerably decreased. Water saturated with carbon dioxide from the fermentation is quite a satisfactory media for some organisms having carbon, oxygen and, of course, water together with an advantageous acid balance (pH). It is therefore important to keep up the sulphur dioxide content of the solution in the trap by re-newing it if necessary.

The next step is to transfer the 300 ml. starter culture to the fermentation vessel and add half a gallon of prepared juice. Take care to be clean in the transfer operation, not allowing drips to run down the outside of the juice vessel, then into the fermentation, as good work can easily be ruined

47

by a contaminating organism. A sterilised funnel will greatly help in performing this operation. It has been found from experience that if only half a gallon of juice is added to the starter, yeast activity is greater in the early stages of fermentation and then after 24 or 48 hours, depending on yeast activity, the rest of the juice can be added. Allow the fermentation to proceed in a warm place (75–80°F.) but try to avoid overheating. A thermometer in the fermentation is a valuable aid to maintaining good temperature control. Fig. 10 shows a fermentation vessel of one gallon capacity fitted with a thermometer and trap. When the fermentation is in progress, attention can be turned to grasping some of the fundamentals underlying the process.

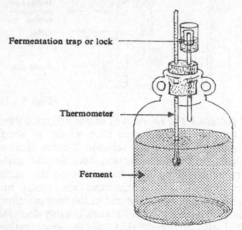

Fermentation trap or lock

Thermometer

Ferment

The theory of fermentation

In the simplest terms, alcoholic fermentation consists of the conversion of sugar to ethyl alcohol and carbon dioxide. Chemically this can be written as follows:

$$C_6H_{12}O_6 \xrightarrow{\text{fermentation}} 2\,CO_2 + 2C_2H_5OH$$

Sugar   carbon dioxide   ethyl alcohol

48

From this certain information can be obtained, namely, that one unit, called a molecule, of sugar under ideal fermentation conditions will produce two molecules of carbon dioxide and two molecules of ethyl alcohol. By knowing a little more chemistry more information can be gleaned. For instance, water can be written H_2O and is made up of two hydrogen atoms and one oxygen atom. The atomic weight of hydrogen is 1 and oxygen 16, thus:

$$H_2 + O = H_2O$$

or $(2 \times 1) + (1 \times 16) = 18$

or 2 gms. hydrogen + 16 gms. of oxygen produces 18 gms. of water.

To apply the same principle to the fermentation equation all that needs to be known is the atomic weight of carbon; this is 12.

Thus: $C_6 H_{12} O_6 \longrightarrow 2 CO_2 + 2C_2 H_5 OH$

$$(6 \times 12) + (12 \times 1) + (6 \times 16) =$$
$$180$$

$$2 \times [12 + (2 \times 16] + 2 \times [(2 \times 12) + (5 \times 1) + 16 + 1]$$
$$88 \qquad\qquad\qquad 92$$

Or 180 gms. of sugar will, if fermentation is complete, produce 88 gms. of carbon dioxide and 92 gms. of ethyl alcohol, theoretically.

In fact, by-products are produced, which we shall see later on proportionately reduce the amount of alcohol formed. It can be said that if the sugar content is known before and after fermentation, the approximate alcohol content can easily be calculated from a knowledge of the equation just discussed and other information regarding the average amount of by-products produced during fermentation. The method of calculation is dealt with in chapter 16. Only the approximate alcohol content can be calculated from a knowledge of sugar content because the process of fermentation is far more complex than so far indicated but for most amateurs this is sufficiently accurate.

More advanced fermentation theory

The yeast cell secretes substances called enzymes. These are substances which act as promoters, making possible reactions which would otherwise be very slow or impossible. Although these substances are present in living yeast cells, the enzymes themselves are purely chemical. *It is the enzymes which are responsible for the alcoholic fermentation of sugar.* Each enzyme is comprised of two parts:

1. **The apo-ferment.** This is a complicated protein substance which determines *the type* of compound to be acted upon.

and

2. **The co-ferment.** This is a relatively simple chemical structure which decides *how* the reaction is to take place.

As an example, the apo-ferment may dictate that fats are to be acted upon and the co-ferment that the action will be oxidation.

The action of enzymes in the alcoholic fermentation

It is generally agreed that basically five enzymes are responsible for the alcoholic fermentation of sugar.

The reactions, greatly simplified, are as follows:
An enzyme called hexokinase transforms glucose/fructose sugar into sugars which are of smaller chemical structure. Actually glucose has six carbon atoms in its "backbone" and is called a hexose, whereas the sugars formed have three carbon atoms in their "backbone" and are called trioses.

One hexose molecule with hexokinase forms two molecules of triose.

The next step is the conversion of triose (three carbon atom "backbone") sugar by a second enzyme, oxidoreductase (or aldolase) into glycerine and glyceric acid. The glycerine remains to give a round and smooth finish to the wine produced.

Thirdly, an enzyme called enolase produces the breakdown of glyceric acid to pyruvic acid. This is largely a simplification reaction; a three carbon atom "backbone" is preserved throughout.

The fourth step is the action of the enzyme carboxylase to form acetaldehyde and carbon dioxide gas from pyruvic acid. This is a breakdown from a three-carbon-atom "backbone" to a two-carbon. Some acetaldehyde remains from this step to add to the flavour profile of the finished wine. Carbon dioxide evolution reminds one of the basic simple formula with which the consideration of alcoholic fermentation started.

The fifth and final step is the reduction of acetaldehyde to ethyl alcohol and this important step is performed by the enzyme zymase.

Allowing the fermentation to overheat or become contaminated with spoilage organisms can lead to the last step becoming an oxidation reaction which then produces acetic acid, otherwise known as vinegar.

Up to the formation of pyruvic acid, phosphate intermediate compounds play a leading role and for this reason phosphate salts have to be present for fermentation to take place at all. Fortunately, sufficient phosphate is usually present in the juice but if fermentation is sluggish a pinch of ammonium phosphate B.P. may well get things going.

For the technically minded, the simplified chemical reactions are set out below:

$$
\begin{array}{ccc}
\text{CHO} & & \\
| & & \\
\text{H-C-OH} & \text{CHO} & \text{CH}_2\text{OH} \\
| & | & | \\
\text{HO-C-H} & \text{H-C-OH} & \text{H-C-OH} \\
| & | & | \\
\text{H-C-OH} & \text{CH}_2\text{OH} & \text{CH}_2\text{OH} \\
| & \text{Triose sugar} & \text{Glycerol} \\
\text{HO-C-H} & & \\
| & (C_3H_6O_3) & (C_3H_3O_3) \\
\text{CH}_2\text{OH} & & \\
\text{Hexose sugar} & & \\
(C_6H_{12}O_6) & &
\end{array}
$$

with arrows labelled: hexokinase enzyme (from Hexose sugar to Triose sugar); oxidoreductase (from Triose sugar to Glycerol)

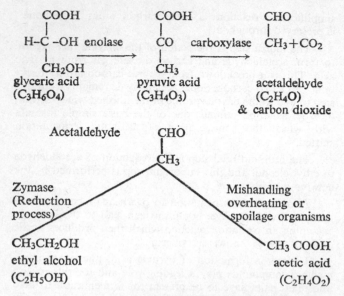

It is now worth while studying briefly the properties of enzymes because they explain many of the practical observations which can be made during a fermentation and give a sound basis from which more detailed study can be made.

The properties of enzymes explained

Enzymes produce heat whilst they are reacting and this is why, although a fermentation may be carried out in a room at 70°F. (21°C.), a fermentation temperature considerably in excess of this may be produced, hence the wisdom of having a thermometer in the ferment.

Enzymes are precipitated by alcohol which means that when a certain alcoholic content is reached, the fermentation will slow down. This can easily be demonstrated in practice and all fermentations exhibit the phenomena.

52

Enzymes are largely insensitive to sterilants and antibiotics. This is the opposite to yeasts which are sensitive to sterilants and antibiotics. A vigorous fermentation is not stopped immediately by the addition of a sterilant even in reasonably high concentration as anyone who has tried with sulphur dioxide will know. It is possible that some free enzyme exists in the solution after the yeast cells have been killed and it is not until the alcohol precipitates the enzyme that fermentation finally comes to a halt.

Enzyme activity

In theory a small amount of enzyme will react with and transform a large mass of reactant substance provided the products are removed and enough time is given. In practice this is nearly impossible. During the alcoholic fermentation of sugar, alcohol and other substances are produced and remain to form the wine—a subsequent slowing down of the reaction therefore takes place due to enzyme precipitation.

Generally most enzymes have maximum activity between 86–104°F. (30–40°C.). This temperature is too high for yeast activity as yeast cells *producing* the enzyme cease to function happily at above 82°F. and are killed at 104°F., and this is the reason why fermentation temperature should NOT EXCEED 82°F (28° C).

Other products of fermentation

In addition to alcohol, we have already seen, glycerol is produced and that this brings a smoothness to the wine flavour. The actual percentage varies with wine type and fermentation conditions and it is worth noting that sulphur dioxide in large doses will adversely affect the balance of alcohol and glycerol produced. To explain this it is necessary to refer to the simplified fermentation reactions on page. 52 From these reactions it will be seen that a substance called acetaldehyde is formed and under the proper conditions of fermentation this is then converted to ethyl alcohol by the enzyme zymase.

Acetaldehyde in common with other aldehydes forms with sodium bisulphite or sodium metabisulphite a substance called a bisulphite compound, and the reaction is shown below:

$$CH_3C\!\!\!\overset{O}{\underset{H}{\diagup}} + NaHSO_3 \longrightarrow CH_3 - C\!\!\!\overset{H}{\underset{SO_3H}{\diagup}}\!\!OH$$

acetaldehyde   sodium   acetaldehyde
                bisulphite   bisulphite compound

The importance of this is if acetaldehyde is bound up in this manner it cannot be converted by zymase into ethyl alcohol. Very large concentrations of sulphite will in fact produce glycerol as the principal product and alcohol as a subsidiary due to the fermentation being largely stopped from proceeding beyond the glycerol stage. With the relatively small amounts of sodium metabisulphite recommended in other chapters to inhibit browning of white wines and give bacterial resistance to any wine, the amount of acetaldehyde converted is very small but if large doses are added *prior* to fermentation the yield of ethyl alcohol is reduced and a proportionately larger amount of glycerol produced instead. A deliberate attempt to produce extra glycerol in a wine by this method is wrong as the whole character of the fermentation will be affected.

The addition of sodium metabisulphite *after* fermentation only gives better antioxidant and bactericidal properties—it does not give higher glycerol content.

Succinic acid is also produced during fermentation and the amount formed is independent of the fermentation conditions prevailing. Advantage is taken of this in fermentation investigations where samples are withdrawn and analysed for succinic acid to establish the stage the fermentation has reached.

Aldehydes in addition to acetaldehyde are produced as are alcohols other than ethyl alcohol. In simple terms an aldehyde is the halfway house between an alcohol and an organic acid and is formed by oxidation of an alcohol.

54

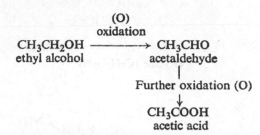

$$CH_3CH_2OH \xrightarrow[\text{oxidation}]{(O)} CH_3CHO$$

ethyl alcohol → acetaldehyde

Further oxidation (O)

$$CH_3COOH$$
acetic acid

We have seen how acetic acid can be formed by the mismanagement of a fermentation. In practice a very small amount of acetic acid is always formed. Very small amounts of other more complex acids are also formed by oxidation and other reactions.

When acids combine with alcohols substances called esters are formed; these contribute to the bouquet of the wine although to a lesser extent than once thought. An example is ethyl alcohol combining with acetic acid to form ethyl acetate which, when pure, has a characteristically "pear drop" aroma. In a wine more than a very small amount of this ester produces volatile acidity; this is most undesirable.

$$CH_3\,CH_2\text{-}\!\left(\overline{OH+H}\right)\!\text{-}OOCCH_3 \rightarrow CH_3\,CH_2\,OOCCH_3 + H_2O$$

ethyl alcohol   acetic acid   ethyl acetate—an ester

Acids containing nitrogen, called amino acids, are produced during fermentation, probably from the breakdown of complex proteins present in the fruit juice. Modern methods of analysis including gas chromatography—the same technique used to analyse a blood sample taken from a driver suspected of being over the 80 milligramme limit—have shown that these amino acids either alone or in combination with the other substances produced during fermentation are a large factor in producing the bouquet of a wine. Two amino acid formulae are given below. Glycine is the most simple amino acid known and the other one tryptophane is given, not to confuse but show that very much more complicated structures do exist yet retain the same characterising groups, viz.—COOH organic acid and NH_2 amine.

$$NH_2 \cdot CH_2COOH$$

Glycine ($C_2H_5O_2N$)

TRYPTOPHANE ($C_{11}H_{12}O_2N_2$)

Condition for yeast activity

Temperature is an important consideration and a fermentation should not be allowed to exceed 82°F. (28°C.) which should be regarded as the maximum for working purposes—the reason for this is explained under enzymes. Although yeast will withstand lower temperatures, do not allow a fermentation to drop below 66°F. (18°C.) for optimum results. Fluctuations in temperature also affect yeast activity.

Ultra violet light reduces the activity of yeast cells and can, if the exposure is prolonged, kill the yeast. For this reason, fermentations should not be placed on a window-sill in direct sunlight and restricted light is to be preferred if it can be arranged and satisfy the requirements of temperature.

Having learnt something of the theory underlying the fermentation process, we can return to the more practical aspects.

As the fermentation reaches its final stages, yeast activity will decrease due to enzyme precipitation and because th products of fermentation remain and are not removed. A decision now has to be made. Is a slightly sweet, downright sweet or completely dry wine required? This depends to some extent on the starting conditions. Having consideration of the starting specific gravities recommended in chapter 3 the following specific gravities indicate the *finishing* point to produce the corresponding wine style after further processing.

Style	Finishing S.G.
Dry red wines	Ferment to dryness
Dry white wines	Ferment to dryness
Sweet white table wines	1.020
Medium sweet dessert	1.010
Sweet dessert	1.038

Take specific gravity measurements during the latter stages of the fermentation, taking care to use a spotlessly clean hydrometer and cylinder, to determine the finishing point corresponding to the style required.

At the desired termination point of the fermentation, remove the bung fitted with thermometer and trap and substitute a sterile bung. If possible place the vessel in a domestic refrigerator as near as possible to the freezer unit, set the refrigerator to the middle position of its scale and leave the jar there for three days.

At this stage the next step of racking, stabilising and fining has to be anticipated and the necessary equipment and materials collected together. Additionally, if a sweet or semi-sweet table wine is being made some treatment will be necessary to prevent further fermentation. A true Dessert style wine will require fortification. These and other operations are dealt with in the next chapter.

Do's and Don'ts

DO continue to work cleanly.

DO decide what you intend making before fermentation.

DO try to understand the principles underlying fermentation as a greater understanding of what is going on is achieved in this manner.

DO NOT despair if the chemistry of fermentation seems forbidding, just read the explanation a little more slowly and forget the chemical formulae.

DO NOT ferment carelessly. A little effort produces good results.

DO NOT think the fermentation process produces ethyl alcohol and carbon dioxide only: other important side reactions also occur.

DO NOT allow the temperature to drop below 66°F. (18°C.) or exceed 82°F. (28°C.).

CHAPTER 6

Racking: sweetening and fortifying

Racking

THIS operation is the removal of clear wine from the settled deposit of yeast which is referred to as "the lees."

When the fermentation has ceased or been arrested by refrigeration the wine should be removed without delay from the yeast deposit. This will prevent the wine acquiring the rather unpleasant flavour of autolysed yeast.

If residual sweetness exists and the wine is not to be fortified add 1.0 gms. per gallon of potassium sorbate. This substance is used in France, Germany and the U.S.A. and, since Aug. 1974, in the United Kingdom for wines, as well as for certain other foods. Any laboratory supplier can provide supplies of potassium sorbate and for use with wine the purest grade sold should be purchased*, and the amount given above not exceeded.

There is no advantage in racking during fermentation unless "sticking" has occurred, then the action of racking will dissolve some air in the wine which may help it to start fermenting again. Even this practice is somewhat dubious because a good stirring of the fermentation with a clean, sterilised stirring rod will produce the same effect *without* racking.

* Marketed as Sorbistat K. by
Rogers (Mead) Ltd., 27 Vicarage Road Wednesfield, Staffs.

Racking serves a second purpose. During the course of fermentation, gases other than carbon dioxide are sometimes produced and the action of moving the wine from the fermentation vessel by racking often helps to remove such gases. Racking at a later stage also assists maturation by giving the wine contact with air to start essential oxidation changes and this facet will be dealt with in a later chapter on Ageing and Maturing.

For racking a syphon tube is necessary and fig. 11 shows two types.

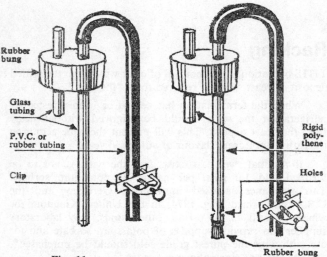

Fig. 11

A. can be of glass and plastic. Many people use this pattern but care is necessary as these are made of glass and therefore are rather delicate.

B. This pattern is made by taking a piece of rigid polythene tubing and cutting it to the desired length. Sealing one end with a small rubber bung and 12 mm. above this, drilling two holes opposite each other in the tubing. A piece of rubber or P.V.C. tubing of sufficient length to reach the bottom of the fresh vessel fitted with a Mohr spring clip, is attached to the top of the polythene tube.

Both patterns require a carrier cork or bung of suitable size to fit the fermentation vessel which is also provided with a vent tube. The vent tube can be of plastic or glass and to prevent entry of extraneous material a small loose plug of cotton wool may be inserted and moistened with three or four drops of 1% sodium metabisulphite solution. The technique of racking is to insert the racking tube into half the depth of the fermented liquid to be transferred. Release the clip and by gentle blowing through the vent tube the syphon will be made and liquid transference to a clean sterile jar can commence. The assembly is shown in fig. 12.

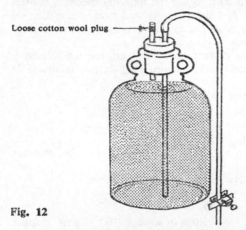

Loose cotton wool plug

Fig. 12

As the liquid level falls, carefully insert the racking tube further into the wine with the transfer tube at the bottom of the fresh vessel; as wine enters air is expelled and contact with air is minimised. As soon as the first lees enter the bottom of the racking tube, release the spring clip to stop the flow. Some loss is inevitable as the lees have to be thrown away and only justify special treatment when a large volume is involved. Even so, the extra wine obtained has to be carefully blended with the bulk due to its lower quality.

The clear wine is now ready for sweetening if necessary, fortification if required or if not properly bright, further settling and fining.

Sweetening

We have already seen that different styles of wine require different terminal specific gravities to obtain a dry, semi-sweet or sweet wine. But even with careful control what was intended to be a sweet wine can easily ferment right through to dryness. To obtain a sweet wine under these circumstances, sugar will have to be added.

Before adding sugar, *always* rack the wine from the yeast lees.

Some recipes give the exact amount of sugar to be added but for those who wish to use the terminal specific gravities given in Chapter 5, 1 lb. of granulated sugar per gallon or 100 gms. per litre raises the specific gravity by approximately 0.036 or 36 degrees.

To obtain the correct sugar addition, use the following formulae and round off to the nearest $\frac{1}{4}$ lb. per gallon or 25 gms. per litre.

The alternative calculations are:—

$$\frac{Degrees\ S.G.\ desired - Degrees\ S.G.\ determined}{36} \times$$

$$\text{gals of wine} = \text{lbs. of sugar}$$

or

$$\frac{Degrees\ S.G.\ desired - Degrees\ S.G.\ determined}{36} \times$$

$$\text{litres of wine} \times 100 = \text{gms of sugar}$$

Thus, if there are 10 gallons of wine with a specific gravity of 1.020 and a specific gravity of 1.038 is required proceed as follows:—

$$\text{lbs. of sugar required} = \frac{38 - 20}{36} \times 10$$

$$= \frac{18}{36} \times 10$$

$$= 5 \text{ lbs.}$$

The sugar can either be added dry and well stirred in or melted in a little warmed wine.

Fortification

Dessert wine fortification

In this type of wine alcohol is added either to arrest fermentation and leave some unfermented sugar and/or give good keeping properties.

Selection of spirit

Generally speaking a neutral flavoured spirit such as Vodka is desirable for fortification as it will leave the flavour profile of the wine unchanged. However, a carefully blended combination of Cognac and wine provides an alternative when a Port style wine is being attempted.

Method of fortification

To fortify a dessert style wine using 40% (70° proof) spirit the St. Andrews Cross method may be used. For example, say the wine to be fortified had a strength of 12% lacohol and it is to be increased to 20% alcohol. This can be written as follows:

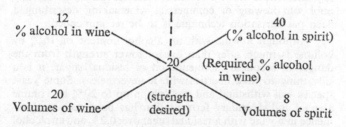

If at the middle of the cross is written the desired % alcohol content, at the top left the % alcohol present in the wine and at the top right the % alcohol in the spirit. The top left figure is subtracted from the one in the centre and the answer written at the bottom right. The top right figure is treated in the same way but the fact that the answer is a negative quantity ignored. In this way the blend of wine/spirit to obtain 20% by volume spirit is obtained. The dotted line is to show wine alcohol content and volume of wine in the blend is on the left and spirit % and volume in the blend on the right. In this case 20 volumes of wine require 8 volumes of spirit to give a wine with 20% alcohol.

The usual level of fortification is up to 18% for dry or medium sweet aperitif or dessert wines and 20% for sweet dessert wines.

Fortification for preservation of unfermented sugar where potassium sorbate is not used

We have seen in chapter 5 that the fermentation enzymes and yeast cells may be inactivated by heat. This is beyond the scope of the amateur as, to be effective, and yet not damage the flavour profile of the beverage, the heating must be for a carefully controlled duration at a critical temperature. It can be done but the apparatus is costly and delicate, involving glass spirals immersed in hot and cold water and a carefully regulated wine flow. For those keen to know more, any book on brewing or commercial winemaking describing a flash pasteurisation technique is to be recommended.

Although a wine with an alcohol content of 16% by volume is much safer than one of lower strength from the danger of secondary fermentation of residual sugar, it is a silly thing to take liberties with it nevertheless. Some yeast species will withstand and multiply in up to 20% by volume alcohol and secondary fermentation has been seen to commence in a wine with a residual sugar over 0.2% and an alcohol content of 17.5% by volume all too easily due to careless handling. The obvious thing to do is continue to take care and use clean sterile equipment.

For practical purposes, if clean techniques have been used, an alcohol content of 16% will assure a reasonably good keeping quality although a high proportion of unfermented sugar may still be present. Careless handling and contamination could lead to secondary fermentation problems.

To fortify, the procedure is to determine the alcohol content of the wine *before* sweetening (using the pre- and postfermenatiton specific gravities and the calculation given in chapter 16.)

The addition of sugar causes an expansion in volume of approximately 6¼% for every 1 lb. sugar per gallon added and a correction for this can be obtained by measuring the volume of *unsweetened* wine after the alcohol content has been determined and again after the addition of the sugar. The difference is the volume expansion. Two sums now need to be done.

1. Using the St. Andrews Cross method, first the alcohol required for the volume of wine *before* sweetening is calculated. The method can best be shown by an example. Say we have 1.5 gallons of wine of 11.5% alcohol before sweetening and 1.9 gallons after sweetening to be made to 16% alcohol with 40% (70° proof) spirit. Thus taking the unsweetened wine first and using the same method as previously:

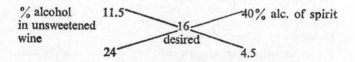

% alcohol in unsweetened wine — 11.5 · · · 40% alc. of spirit — 16 desired — 24 · · · 4.5

24 volumes of wine requires 4.5 volumes of spirit.

1.5 volumes of wine require $\dfrac{1.5}{24} \times 4.5 = 0.28$ vols of 40% alcohol spirit

OR 1.5 gallons unsweetened wine needs 0.28 gallons of spirit.

2. But we had a volume expansion of 1.9 — 1.5 = 0.4 gallons and theoretically this contains *no* alcohol.

So using the same method of calculation to determine

the volume of alcohol necessary to raise the strength of 0.4 gallons of water to 16% alcohol we write:

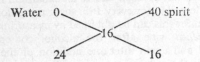

Water 0⟍   ⟋40 spirit
          ⟍ 16 ⟋
          ⟋   ⟍
        24⟋   ⟍16

24 volumes of water require 16 volumes of 40% alcohol spirit.

0.4 volumes of water require $\dfrac{0.4 \times 16}{24} = 0.27$ volumes of spirit

The total 40% spirit which requires to be added to the 1.9 gallons of wine is 0.28 plus 0.27 = 0.55 gallons or approximately half a gallon of spirit.

Note that the expansion requires almost as much spirit as the unsweetened wine because theoretically the volume expansion contains no alcohol. The table on page 234 gives comparisons of proof strength with % alcohol content by volume. To use a fortifying spirit of higher strength simply look up its % v/v of alcohol and use it instead of 40 in the St. Andrew's Cross method.

Alternatively, say a spirit of 92° proof spirit is being used, that is to say, Polish or Russian Vodka, reference to the table on page 234 will show this corresponds to an alcohol content of 52.5% v/v. So if we have one pint of wine at 12% v/v alcohol and it is to be made to 16% v/v alcohol, proceed as follows:

Write the % alcohol content of the Vodka at the top right of the cross and the % alcohol by volume in the wine at the top left. The desired alcohol content is written in the middle.

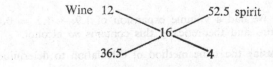

Wine 12⟍   ⟋52.5 spirit
          ⟍ 16 ⟋
          ⟋   ⟍
      36.5⟋   ⟍4

We thus see that 36.5 pints of wine would require four pints of spirit and as we have only 1 pint of wine of Vodka will be required.

$$\frac{4 \text{ pints}}{36.5}$$

Litre bottles are now in common use and the following table has been calculated to simplify fortifying by up tp 10% by volume any wine with a strength between 12 & 16% by volume, using spirit of either of the strengths shown. The table is not intended to replace the calculations already described as some approximation is involved.

TABLE 3

Proof Strength.	% by Volume	1	2	3	4	5	6	7	8	9	10mls
					Percentage increase.						
65.5	37.51	45	94	148	207	273	347	430	525	633	757mls
70.0	40.07	40	84	131	183	239	302	371	448	534	632mls
85,0	48.62	30	62	95	131	170	211	255	302	353	409mls
90.0	51.46	28	57	87	120	155	191	231	273	318	366mls

To use the table to increase the alcohol of a wine between 12 & 16% by volume proceed as follows. First decide the strength to which the wine is to be fortified, next the alcohol content of the spirit which is to be used. Subtract the strength of the wine from the desired strength and then determine the volume of spirit required per litre from the table. For example, if a wine of 13% by volume alcohol is to be fortified to 20% using 90° proof spirit, 20—13 = 7% and then looking in the table for 7% increase with 90° spirit it will be found 231 mls. of spirit is required for each litre of wine.

Fortifications can after a period of days cause a slight haziness to appear in the wine. The reasons for this can be either:

1. Substances which are soluble in the weaker alcoholic strength of the wine *before* fortification have by the fortification either been rendered less soluble or insoluble. This change can take some time to evidence itself and no definite statements can be made regarding individual wines, each case being largely a matter of experience.

67

This is why it is important to fortify *before* fining and aging, the respective processes for which will be described later

or

2. The denaturing of protein material present in the wine by the fortifying spirit. Denaturing is a non-reversible process which *converts* protein from one chemical state to another, in this case from water soluble to water/spirit insoluble. The easiest example from everyday life is the frying of an egg: the albumen is a pale straw transparent liquid when the egg is freshly broken but when it gets into the hot fat of a frying pan, it turns to the familiar white. Nothing will change the white solid back to the liquid transparent albumen; the protein has been denatured. The protein matter present in wine (or juice) is affected in a similar way, producing an insoluble deposit when high strength alcohol is added. A fining agent can be used to remove protein haze and here again we see why it is necessary to fortify *before* fining.

After fortifying, place jars which are small enough in the domestic refrigerator for a week. In this way the formation of any haze will be accelerated, most substances being less soluble in the cold than at room temperature. The wine can then be racked into fresh storage bottles and left to mature.

Do's and Don'ts

DO approach each operation knowing what it is about—it will help to get good results.

DO keep apparatus simple and clean.

DO rack before either sweetening or fortifying.

DO use neutral flavour-spirit for fortification unless special character is to be imparted by the spirit.

DO refrigerate after fortification—it helps speed precipitation.

DO NOT despair if a wine intended to have residual sugar ferments right out—it can be sweetened to the level desired.

DO NOT sweeten the wine whilst it is still on the lees.

DO NOT waste time processing small amounts of lees, the clear wine is much more important.

DO NOT automatically buy racking apparatus—you can probably make it more cheaply.

DO NOT fortify the wine whilst it is still on the lees.

DO NOT be put off by a little arithmetic—it is only simple.

CHAPTER 7

Aging

THIS process is probably one of the most elusive to explain in terms which are easily understood, yet it is an essential requirement for every wine. When a newly fermented wine is racked off its fermentation lees, it is, regardless of its colour or style, rather harsh, tends to be yeasty in aroma and flavour and lacks the roundness and balance which develops on storage. The aging time will vary from wine to wine—some characteristics developing whilst the wine is stored in bulk and others whilst it is in bottle. To understand the process better let us consider the theory underlying the aging process.

The mechanism of aging

One thing that can be said for certain is that during the time a wine is in storage slow oxidation takes place, altering the characteristics of the wine progressively. The rate of oxidation varies with the storage condition, wine in cask oxidising more rapidly than that in bottle and light white wines much more quickly than full red wines.

In the chapter on fermentation we saw that the fermentation process produces not only alcohols, other than ethyl alcohol, but acids and substances half way between an alcohol and an acid, called aldehydes. The marrying together of alcohols and acids produces a further group of substances called esters and it is these substances which provide some of the delicate aroma apparent in a well-aged wine. Aldehydes are rather sharp to the nasal senses and for over a hundred years various workers have provided evidence to show that slow oxidation progressively changes these substances into their corresponding acids which, in turn, over a further period of time, combine with the free alcohols present to produce esters. Thus over a period of time "newness" is lost and fragrant signs of maturity establish themselves. The oxidation of alcohols produced as by-products of fermentation and tannins from the original fruit also add to flavour development and bouquet. The acids involved are not confined to simple chemical structures but extend to the amino acids already mentioned in chapter 5.

This is only part of the story because if oxidation is allowed to proceed too fast, by excessively aerating the wine, or too long by not correctly gauging the point (by tasting, nosing and observing) when a wine has reached its zenith, a decline in quality will occur. It is for this reason that experience is always helpful in establishing the point after which a close watch has to be maintained to arrive at the optimum time for bottling.

White wines

Excessive oxidation of a white wine, especially one which has a rather delicate character, will cause the onset of premature browning which is accompanied by a dulling of the flavour and if allowed to continue for long enough a "madeirised" odour and taste will be assumed. When this stage is reached the wine is only marginally useful for cooking. From this it will be obvious to the reader that racking, although necessary, should be carried out in the manner recommended in chapter 6 and that during aging vessels should be kept topped up to 3 mm. from the bung or cork to minimise contact with air. Shaking of storage vessels, especially after opening for inspection, should be avoided lest more air than normal will be dissolved in the wine.

Red wines

Generally speaking one can expect a red wine to deposit some colouring matter and tannin during the period it is aging. This is brought about by slow oxidation of soluble pigments and tannin compounds to form insoluble matter which is then deposited. This sedimentation is slow and in a wine with a lot of colouring matter and/or tannin will extend to the period after bottling because during the racking and bottling operations sufficient oxygen is dissolved by the wine to allow slow maturation to continue. Aldehyde production and oxidation of other constituents contributes to the maturation process.

The effect of sulphur dioxide on maturation

Red wines

For red wines the addition of sulphur dioxide at the aging stage cannot be recommended as anything other than a remedial step to prevent bacteria spoiling a wine after some carelessness in handling.

Dessert wines

Do not benefit from sulphur dioxide unless the alcohol content is low (below 18% by volume) and bacterial growth is feared; this latter condition would be indicated if a haze accompanied by an off taste appeared. See also chapters 12 and 16.

White wines

Excessive amounts of sulphur dioxide are objectionable to the nose and the palate but a small amount added (to white wines in particular) has a beneficial effect.

The addition of sulphur dioxide to a white wine reduces the speed of oxidation and preserves the light colour of the wine. The reason for this is that dissolved oxygen is used up in converting sulphur dioxide to sulphur trioxide, which in turn dissolves in water to give a minute amount of sulphuric acid. The reactions are as follows:

$$2 SO_2 + O_2 \longrightarrow 2SO_3$$
Sulphur dioxide oxygen$\longrightarrow$sulphur trioxide
$$SO_3 + H_2O \longrightarrow H_2 SO_4$$
sulphur trioxide water$\longrightarrow$sulphuric acid

The production of small amounts of sulphuric acid does affect the wine's total acidity slightly and for this reason the addition of sulphur dioxide should not be too great and a dose of 50 ppm prior to ageing is recommended. This corresponds approximately to the addition of ½ gm. of sodium metabisulphite per gallon of wine.

To make the addition, remove some wine and dissolve the sodium metabisulphite, then add this to the bulk of the wine, fill the vessel up and mix in with a clean glass or plastic rod. Do not whisk or beat air into the wine and complete the operation speedily. Finally close the vessel with a bung.

Effect of sunlight

Ample evidence exists to show that sunlight falling on bottles of maturing wine rapidly increases aging and wine should not be stored in the direct light of the sun for this reason. Ultra violet light has been used in some parts of the world to age wine artificially but due to the lack of predictability, its use has not been widely favoured.

Try to store wines in the dark—failing this, in a place where light is small and direct sunlight does not enter.

Aging temperature

Table wines age best at a temperature maintained as evenly as possible between 50 — 60°F. (10°C. — 15.5°C.). In this way oxidation changes are progressive and some degree of prediction can be made as to when the wine will be at its best for bottling. Dessert wines with their higher alcohol content can tolerate and often benefit by slightly higher temperatures, 60–70°F. (15.5°C.–21°C.) generally giving satisfactory results, but 70°F. (21°C.) should be regarded as the maximum. Over this temperature oxidation tends to become faster than is desirable, producing a darkening of the colour and a flatness to the taste.

Choice of vessel for aging wine

If a wine is racked twice during a year with the sterility precautions already outlined in chapter 6, sufficient oxygen will dissolve to account for the necessary oxidative changes to assist maturation. Glass and earthenware vessels cleaned in a proper manner are therefore perfectly satisfactory. Most of the textbooks on wine refer to aging carried out in wooden casks but there are certain disadvantages.

For a very long time it has been known that air gradually diffuses through the wood and subsequently dissolves in the enclosed wine. In similar fashion water and alcohol from the wine diffuse through the wood to evaporate into the surrounding air and this leaves a slight air space or ullage above the wine. For this latter reason casks have to be inspected at regular intervals and topped up to prevent the wine dissolving excessive amounts of oxygen from the ullage and starting to acetify by the oxidation of actaldehyde to acetic acid. Careful attention is called for to ensure that the casks are topped up and do not leak. This creates the added problem of having odd amounts of wine available for the purpose, and this together with the difficulties of coopering (making sure the cask is constructed correctly), cleaning and sterilising, means casks are not ideal for amateur use.

Wines stored in polythene containers oxidise more rapidly with consequent loss of quality than identical wine stored in glass. Low density polythene cannot be recommended for storage over a period exceeding 2–3 months whilst high density polythene is usually satisfactory for 6–9 months.

Aging in bottle will be dealt with in the chapter dealing with bottling.

When the right degree of aging has been achieved, the wine has to be clarified and this leads us to the next step in the sequence.

Do's and Don'ts

DO make sure containers are kept topped up with wine.

DO use a dose of sulphur dioxide during the aging of white wines.

DO try to keep the aging temperature steady.

DO rack the wine off any sediment formed twice during a year—it will assist aging.

DO NOT store wine in partially filled containers.

DO NOT shake or disturb wines which are aging.

DO NOT age wines in sunlight.

DO NOT allow the temperature to exceed 70°F. (21°C. during aging and for table wines try to keep below 60°F. (15.5°C.).

DO NOT use casks without carefully considering the disadvantages.

DO NOT use polythene vessels for long term storage.

CHAPTER 8

Filtration and fining

WHEN the aging of a wine is complete, the best course of action is to rack it into a fresh clean vessel and, under ideal conditions, leave all traces of sediment behind. If this is possible the wine will be "star bright"—a term used to indicate it is perfectly clear with no suggestion of haze whatsoever. With care this is possible on a surprisingly large number of occasions. However, it may be found that despite careful racking the wine is still cloudy or possesses a haze. If possible the cause of the haze should be determined by applying the tests given in chapter 16.

Two courses of action are then open to the winemaker

1. Filter the wine

    or

2. Treat it with a suitable fining agent to deposit the haze and leave the wine clear and bright.

Before deciding which course of action to take it is worthwhile finding out if the haze is simply suspended solid matter or if it is of a colloidal nature. This can easily be done by filtering a glass full of the wine through a folded Whatman No. 1 filter paper and if after filtration the wine is *not* star bright the haze is likely to be colloidal and it is best to apply the fining tests given later in the chapter to decide which fining agent is most suitable for the purpose. Filtration is not successful for the removal of colloidal hazes and is best suited to remove suspended solids. Filtration is, of course, practised widely in commerce but there are generally large volumes of liquid being handled and special techniques used. In the hands of the amateur, because of the restrictions encountered with relatively small volumes, wine is easily given a quite undesirable taint—usually of asbestos—or robbed of some of the flavour and bouquet highlights. For those wishing to use filtration either as a primary means of clearing and polishing, or simply to decrease the loss of wine after fining, the various methods available, together with the restrictions and precautions necessary, are given below.

1. Filtration by gravity using a filter paper

This method is useless for the removal of a fine haze but is, however, useful for removing "fliers" left behind after fining.

A conical shaped funnel is used together with a filter paper circle of suitable particle retention grading and size. A Whatman No. 1 grade is the finest grade that can be recommended for amateur winemakers' use and many coarser grades than this will, from individual experience, be found satisfactory. In fact either Whatman G.P. or No. 4 will usually prove satisfactory. Finer grades than Whatman No. 1 require quite long periods for filtration to take place and the filters clog relatively quickly.

Pick the largest funnel available and fold the filter pape as shown in either fig. 13 or fig. 14.

The following is one of the simplest ways of folding a filter paper.

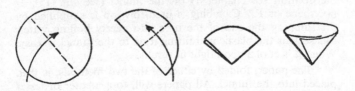

Fig. 13 How to fold a filter paper

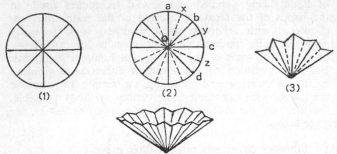

Fig. 14 How to flute a filter paper

How to flute a filter paper

First make four folds in the paper so that the latter is divided into eight equal sectors (1), the two halves of the paper on each occasion being folded forwards, so that all the folds tend to be concave. Now take each segment in turn (e.g. *aob*, (2)) and fold the points *a b* backwards until they meet, so that a new convex fold *ox* is made between them: continue in this way making new folds *oy*, *oz* etc., around the paper.

The method shown in fig. 13 does not give such a fast filtration as that of fig. 14 because the filter paper is twice as thick on one side as the other (as shown in the diagram) and therefore filtration is slower on that side.

To prevent the wine cascading through air and dissolving oxygen, which could cause premature browning in the case of white wines, place a piece of sterilised tubing (for Method of Sterilisation see chapter 1) on the funnel (see fig. 15). If polythene or P.V.C. tubing is used this job is simplified by first dipping the end of the tube into nearly boiling water, this softens the plastic and fitting it on to the funnel is easy. When it is cool a good tight fit results.

The paper, folded by either of the two methods, is then placed into the funnel. All papers will, to a greater or lesser extent, impart a taint to a wine. To prevent this, wash the paper with boiling water several times and then let it drain, covering the funnel with an inverted saucer or clockglass to prevent any foreign matter getting into the filter.

Fill the paper with wine to within 3 mm. of the top, cover with the inverted saucer and leave to filter, topping up as necessary. Remember to cut a small groove in the side of the bung to allow the escape of air as the vessel fills. The funnel is supported by the cork and a stand is unnecessary.

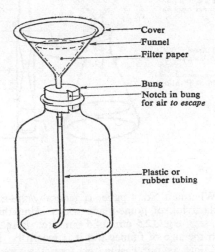

Fig. 15

The disadvantage of this method is that it is slow and some oxidation invariably occurs.

2. Filtration using a suction pump

The use of this method accelerates the speed of filtration and allows the use of finer filters. However, it requires a simple suction pump and a funnel and flask of special design to withstand a partial vacuum. The biggest objection is some of the wine's bouquet is nearly always lost and if considerable care is not exercised water can suck back from the pump into the filtered wine. When assembled, the apparatus appears as in fig. 16. All the items are readily available from a supplier of laboratory apparatus.

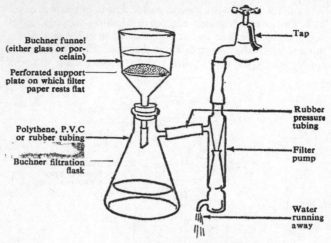

Buchner funnel (either glass or porcelain)

Perforated support plate on which filter paper rests flat

Polythene, P.V.C or rubber tubing

Buchner filtration flask

Tap

Rubber pressure tubing

Filter pump

Water running away

Fig. 16

Whatman No. 4 paper is generally the most suitable for the clarifying of home-made wines using this method. The circle size, e.g. 12.5 cm. 14.5 cm. etc., is selected to exactly match the size of funnel and it is worth buying these at the same time as purchasing the funnel to make sure of the proper size.

After positioning the filter paper, in the funnel, it is moistened with a little water to bed it down on to the support plate.

The water is turned on at the tap and wine poured gently into the centre of the filter paper to almost fill the funnel. Wine will be sucked through into the flask below, and it is possible to keep replenishing the funnel until the flask is nearly four fifths full when the wine will have to be removed to a storage vessel. To halt the filtration, allow the funnel to nearly empty and then gently slip the rubber pressure tubing off the side arm of the flask. Air will rush into the flask to compensate the outside and inside pressures. The water at the pump now can be turned off, the funnel removed from the flask and the wine transferred to a storage vessel. It is essential the water is *not turned off before the suction tube is removed* or water will suck back into the wine. A simple way to prevent this happening is to buy or make a trap as shown in fig. 17. This must also be of thick walled glass to prevent implosion (a breaking inwards) of the walls. With a trap installed between the pump and the flask, filtration can be stopped by simply turning the water off and any sucked back water is trapped and cannot now reach the wine.

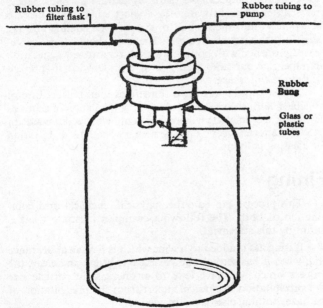

Rubber tubing to filter flask

Rubber tubing to pump

Rubber Bung

Glass or plastic tubes

Fig. 17

Proprietary filters which utilise filter pads are also on the market. Essentially these filters consist of a support for the filter pad so that it will not break during use, and an upper and lower housing which serve to conduct the wine onto the

81

filter and afterward away from it. Some filters of this type utilise a principle which consists of first coating a coarse filter pad with kieselguhr powder which then actually does the filtration. Examples of these filters are shown in plate No. Fig. 17.

Cellulose or asbestos pulp as a filter medium

The use of cellulose pulp is sometimes recommended instead of filter sheets or pads and can be quite successful, providing precautions are taken to prevent interference with the flavour of the wine.

Cellulose pulp can be made by tearing up filter papers and soaking them in water for a short while. The slurry is then poured into the funnel to give an even coating and the excess water sucked through. The mat should then be repeatedly washed with boiling water to remove any suggestion of paper flavour and finally with a little wine to prevent dilution taking place.

Asbestos pulp was once sometimes used but because of possible health risks associated with the inhalation of asbestos its use by the amateur winemaker cannot be recommended.

The alternative or complementary method to filtration for clearing is fining.

Fining

This process can be either physical, chemical or a combination of both. The following examples illustrate what is meant by this statement.

If isinglass is added to a wine which is hazy in appearance and, when it has settled, the beverage is bright and clear, the process which will have lead to an increase in particle size and precipitation is *physical*; apart from the precipitation of the haze, nothing else has happened.

Should the same wine contain in addition to the haze an excess of iron as shown only by a chemical test and it is blue fined (this will be discussed later) iron will be removed and deposited by a *chemical* process; the haze however might remain due to it being derived from another origin.

Lastly, if the same wine, which we have said is hazy and contains excess iron is first blue fined and then fined with isinglass, *chemical and physical* fining will have taken place.

Putting the matter in its simplest terms, fining can be for two purposes, clarification and/or purification.

Theory of clarification by fining

It was at one time thought that when a fining was added to clarify a wine, it did so by the molecules of the fining agent forming a mesh-like structure which settled down through the wine, trapping the colloidal haze molecules and taking them to the bottom because of the increased size and weight. Ingenious as it is, this theory has now definitely been proved to be false.

About the mid-nineteen fifties a team of scientists working on the continent discovered the true fining action. To understand the theory it is necessary to know that the sub-microscopic particles which make up hazes, called colloidal particles, carry a minute electrical charge. The charge may be either positive or negative depending upon the colloid.

As most people know, if two bar magnets are brought together with either the two North or two South poles facing each other they will repel. If, however, the North pole of one magnet is brought near the South pole of the other, they will attract to each other. This analogy can be applied to a wine with a colloidal haze. All the electrical charges on the particles will be the same and they will therefore repel one another and remain in colloidal suspension. If a substance with the opposite electrical charge on its particles is introduced, cancellation of charges will take place and the particles can amass together to form larger bodies which, with the increase in weight, will then be deposited by the force of gravity. This then is how a fining agent works and because the charges differ depending on the wine haze, so must the fining agent be selected correctly to obtain good clarification.

At this stage it is worth adding a word of caution. It has already been said that the addition of a clarifying fining material to a wine is really the addition of a set of minute electrical charges to cancel those on the colloidal haze particles in the wine. If, therefore, excessive fining is added, not only will the colloidal haze particles be neutralised, but charges due to the fining agent will be introduced and a new haze problem will arise due to the presence of excess fining; it is this which is termed *over-fining*.

Clarifying

How to select the right fining agent

Many of the problems besetting the amateur winemaker centre around hazes due to protein or colouring matter. These consist of positively charged colloidal particles and to remove them from the wine a negatively charged fining is necessary. The most satisfactory of these is *Bentonite*. This is an earth found in America and certain other areas of the world which, because of its special ability to combine with protein material, is often used by commercial winemakers, sometimes after fining with gelatine or egg white, and in this way any slight overfining from the gelatine or egg white is removed.

Whilst generally useful it is nevertheless possible that bentonite may not be the correct fining to use and an alternative will have to be selected to give effective clarification. To avoid a hit or miss method of obtaining the right fining, a simple trial will provide by far the most satisfactory solution.

A simple fining trial

Prepare a 1% solution of bentonite in warm—not boiling—water. (A 1% solution is simply 1 gm. of bentonite powder in 100 ccs. of water). The bentonite will not dissolve but if it is shaken regularly over the period of a whole day and immediately before use, this will be satisfactory.

Also make a 1% solution of gelatine by dissolving 1 gm. of gelatine B.P. in very hot water. Do not boil as this starts decomposition of the gelatine into other products by a process known as hydrolysis. Allow to cool. A fresh solution should be prepared for each trial.

The only other solution required is 1% tannic acid B.P. which can be made by dissolving 1 gm. of tannic acid B.P. in 100 ccs. of water. The procedure is then as follows

Into each of six 125 ml. clear glass bottles, measure 100 ccs. of wine. Using a 5 mls. graduated pipette—this can be purchased from a laboratory supplier—add 0.5, 1.0, 1.5, 2.0, 2.5 and 3.0 ccs. of bentonite to the jars respectively. Shake each jar and label it so that the next day it is clear which is which. On the following day examine the bottles carefully without disturbing them. Look for the one giving a clear bright wine with the smallest bentonite addition. In this way you will not overfine.

Just for example say the 2.5 mls. of 1% bentonite solution in 100 mls. of wine gave excellent results. It is then possible to calculate the addition of bentonite necessary for the bulk of the wine.

Since 1 mls. of 1% solution will contain 1/100 or 0.01 gms. of bentonite

2.5 mls. of 1% solution will contain $2.5 \times 1/100$ or 0.025 gms. of bentonite

and as 1 gallon consists of approximately 4,500 mls., each gallon of wine will require $\dfrac{4500}{100} \times 0.025$ gms. of bentonite

This works out at $45 \times 0.025 = 1.12$ gms. bentonite per gallon of wine.

Going to the nearest quarter of a gram $1\frac{1}{4}$ gms. per gallon of wine.

In simple terms:

Number of grams of fining required per gallon of wine $= 0.45 \times$ the number of mls. of 1% fining most satisfactory in 100 mls. fining trial.

If bentonite is unsuccessful in producing clarity, the next step is to wash the bottles and repeat the trial in exactly the same way but using 1% gelatine. Gelatine removes tannin and because of this it is necessary to add the same amount of 1% tannin to the 100 mls. volumes of wine as gelatine, viz. 1 ml. of gelatine then 1 ml. of 1% tannic acid as well.

Sometimes less tannin than equal proportions will suffice but it is as well to carry out an equal proportion trial to begin with. Establish which addition, e.g. 0.5, 1.0, 1.5 etc, ccs. gives the best clarity and calculate the addition of fining per gallon of wine as explained for Bentonite above. (i.e. $0.45 \times$ No. of mls./100 mls. shown by trial.)

Mixing the fining material into the wine

Finings should always be mixed into a volume of the actual wine which is to be cleared and then this "super concentrated" fining well mixed into the bulk. This method avoids excessive localised fining and distributes the fining throughout the whole body of the wine.

After fining, the wine should be left for about seven days. If it is possible, placing the wine in a refrigerator will considerably speed up precipitation. At the end of seven days if examination shows the lees to have settled well, the wine may be racked off into a clean sterile container using the technique on page 61 of chapter 6.

Types of fining agents used

Many fining substances exist, each possessing its own particular merits and disadvantages, each demanding preparation in a specific manner. To help the reader who not only wishes to learn something of fining in general but who would also like to experiment, the various fining materials in use together with their method of preparation and common dosages are now described. It should be emphasised, however, that a fining trial with the intended fining media in the manner already explained is the most accurate and reliable method of obtaining the correct dosage to use.

Albuminous finings

These are composed of fresh or dried egg whites. They carry a positive charge and whilst being satisfactory for general use do suffer from the disadvantage of easy overfining. Fresh egg whites are used only for relatively large volumes of wine—4–8 whites per 100 gallons being the common dosage recommended. Thus, for the amateur, unless a volume greater than 12 gallons is being processed, a danger of overfining exists.

The technique which seems most suitable is to decide what dose is going to be used and then separate the whites from the yolk of the requisite number of eggs and whisk the white into about 2–4 pints of wine. This is then added with stirring to the bulk of the wine—leave to stand 7–10 days and then rack off from the lees.

Alternatively, using powdered or crystalline albumen, prepare a 1% solution (1 gm. in 100 mls.) in water and conduct a fining trial as already explained. Determine the correct amount to be added to the bulk of the wine, weigh out this quantity of dried albumen, dissolve it in some of the wine and then mix this solution into the bulk.

For the amateur dried albumen is recommended, as it is the easiest to control and overfining is avoided. If excess albuminous fining is added a large protein haze problem will occur from this source.

Bentonite (a Montmorillonite earth with a chemical formula $Al_2O_3 4Si O_2 x H_2O$). It consists of aluminium and silica oxides together with a variable amount of water. As already said this material carries a negative charge and has the ability to combine with and precipitate albuminous and other proteinaceous colloids. A simple fining trial using a 1% dispersion should always be carried out, as already described to determine the correct dose. Weigh out the quantity required for the bulk and mix it into 1–4 pints of wine—depending on the total volume to be treated—shaking frequently over a 24 hour period. Mix the dispersion well into the bulk of the wine and leave to stand for 7–10 days.

This is versatile fining and deserves to be used more by amateur winemakers. Overfining gives an earthy flavour to the wine but is easily avoided by making the simple trial fining.

Carbon (Amorphous or powdered charcoal)

A material which can be obtained industrially from a number of sources and in varying degrees of purity. For wine use a grade which will not contaminate with iron or copper is necessary and should be purchased from a laboratory supplier.

Carbon is used to reduce colour and remove "off" flavours. Unfortunately it requires a very careful trial fining to establish the correct dose or else a near water white liquid with greatly reduced flavour results. Advantage is taken of carbon's properties to remove any trace of colour and undesirable flavour from Vodka.

The details of carbon as a fining agent are included for interest only as its use is not recommended to the amateur who lacks laboratory facilities.

Casein

This amino acid containing material is one of the principal constituents of milk and cheese. It is rather expensive and not simple to prepare. Used under controlled conditions it offers the advantages of a decolouriser without the disadvantages of flavour removal and in this respect is therefore superior to carbon.

A common way to prepare casein is to dissolve $\frac{1}{2}$ gm. of sodium bicarbonate in 100 mls. of water and then mix in 2 gms. of casein. The fining trial is then performed and the calculated quantity of casein added in the form of this 2% solution to the bulk of the wine. The dose required varies between $\frac{1}{4}$–1 gm. per gallon or 12.5–50 mls. of 2% solution. It is common to add approximately the same amount of tannin dissolved in a little wine, e.g. 25 mls. of 2% casein solution and $\frac{1}{2}$ gm. of tannic acid.

This fining is not used very extensively in the United Kingdom but gives good results.

Dried or fresh blood

This used with care is similar in action to albumen but has the disadvantage of easy overfining. In some continental wineries it is still used but its use is declining rapidly. Fresh blood from a slaughterhouse was once favoured and the dose in this case is between 2 and 4 mls./ gallon of wine. Dried blood, used for its ease of storage is added at a rate of approximately ¼–1 gm./gallon of wine.

Not recommended to amateur winemakers.

Isinglass

This must not be confused with the mineral isinglass as it comes from the air sacs of the Sturgeon fish. It is a positively charged fining and therefore similar in its action to gelatine and albumen. Being rather difficult to make into a solution and somewhat temperamental in its action, the use of isinglass has tended to decline in favour of gelatine. However, it is less easy to overfine with isinglass than it is with gelatine although the lees do not compact as well.

To prepare 1% isinglass, soak 1 gm. of the material in 25 mls. of water overnight. Next day add enough wine to make the volume up to 100 mls. and stir well. Carry out a fining trial as for gelatine and use 1% tannin solution in exactly the same way. Leave seven to 10 days to settle and then rack off.

Gelatine

Manufactured from bones and used in glue manufacture, this should be B.P. grade or the purest form available. It is dissolved in hot, *not* boiling, water as decomposition takes place on boiling. A 1% solution is used to carry out a fining trial and because gelatine removes tannin and colouring matter to some extent, it is common practice to add half to the total quantity of tannic acid as gelatine at the time of fining. A fining trial is important because overfining can produce complications such as excessive loss of colour, appearance of fining taste on the palate and haze due to excess gelatine.

The dose varies between ¼ and ¾ gm. of gelatine per gallon of wine together with ¼ to ½ gm./gallon of tannin.

Milk

Containing casein as one of its principal constituents, this provides a much more simple source of the material than preparing a special solution. Used judiciously, it is a useful decolourising agent for wines which are a little darker than required.

Carry out a trial by taking four 100 mls. quantities of the wine to be fined in clear glass bottles all of the same shape and size. Add, using a graduated pipette, $\frac{1}{2}$ ml. of milk to the first, 1 cc. to the second, $1\frac{1}{2}$ ccs. to the third and 2 mils. to the fourth. Give each a shake and allow to stand for 24–48 hours, the one giving the desired colour with the minimum of milk is the one to select. Forty-five × the number of mils. required for the 100 ccs. trial = the volume of milk in mls. to be added to each gallon of wine.

Although not complete, the above list of finings will have served to show that the correct selection of fining material is important.

So far we have dealt with fining for clarification: to conclude the chapter the subject of purification fining will now be summarised briefly.

Purification fining

Because of the many technical difficulties involved, this type of fining is beyond the scope of the amateur winemaker although many will, no doubt, be interested in its application.

Broadly speaking, iron and copper are the most trouble-some contaminants arising in the main from the use of copper, brass, chipped enamel or mild steel apparatus.

Many books on wine science will tell you that the addition of phytate or phosphate salts or tannins will either remove or prevent trouble from iron and copper. Unfortunately experience proves a very expensive teacher and theory does not quite match practice. The situation is redeemed by the fact that a technique known as *Blue fining* in the proper hands will remove iron and copper, rendering a wine more pure and safe from unsightly deposits or unpleasant taint.

Blue fining

This must be done by a skilled chemist who fully understands the process. It consists of adding a very carefully determined amount of potassium ferrocyanide to the wine; this complexes with any iron contamination to form a complex called *Prussian Blue* which is thrown out of solution. With copper a chocolate brown precipitate is formed. In either case potassium takes the place of the contaminating metal. For those who are interested the chemical reactions are as follows:

1. Iron removal

$$3\ K_4\ Fe\ (CN)_6 + 4\ Fe^{+++} \longrightarrow Fe_4\ (Fe\ (CN)_6)_3 + 12K^+$$

potassium ferrocyanide	Ferric ions	Prussian blue	Potassium ions

What this means is that iron is removed in the form of the dye, Prussian blue, and replaced by potassium.

Iron combines with phosphates and tannins in wine to produce deposits; it also gives a metallic taste to wine. Potassium does not do these things.

2. Copper removal

The process for copper removal can be expressed chemically as follows:

$$K_4\ Fe\ (CN)_6 + 2\ Cu^{++} \longrightarrow Cu_2\ Fe\ (CN)_6 + 4K^+$$

Potassium ferrocyanide	Cupric Copper ions	Cupric ferrocyanide (chocolate brown)	Potassium ions

In this case copper is replaced by potassium.

If the process is not carried out by a skilled chemist, a danger of over fining can exist and it is for this reason that Blue fining must not be carried out by the amateur winemaker. Excess potassium ferrocyanide in the acid conditions of a wine could produce traces of potassium cyanide which, of course, is a deadly poison. With proper knowledge of the process, excessive potassium ferrocyanide addition *never* occurs and blue fining is extremely effective. It is practised in many parts of the world, including the U.K.

A patent compound called Cufex is sold in the U.S.A. and has the same action as Blue fining but is much simpler in its application.

Both Blue fining and Cufex require a post fining with gelatine and tannin.

Do's and Don'ts

DO try to rack your wine star-bright if you can.

DO wash your filter thoroughly to avoid possible tainting.

DO *first* carry out a simple trial with Bentonite.

DO progress to another fining if Bentonite is shown to be unsatisfactory.

DO remember overfining is just as bad as not fining. Often the effect is the same.

DO rack the wine from the lees when fining sedimentation is complete.

DO expect some loss of wine by fining—it is inevitable.

DO NOT despair if your wine is not star bright after aging.

DO NOT forget to wash any filter you use free from possible taints.

DO NOT carry out fining blindly. A simple trial will give you a wealth of information.

DO NOT think a fining trial is difficult—it is not.

DO NOT attempt Blue fining. It is not for the amateur.

DO NOT use asbestos powders for filtration—they need specialist techniques.

CHAPTER 9

Bottling

TO many this task will appear routine, a simple matter of pouring the matured wine into a bottle, inserting a cork, covering the bottle top with a decorative capsule and applying an eye catching label. How wrong they are!

Bottling is an important stage for a wine. Done with indifference or carelessness it can ruin all the effort which has been put in during earlier stages; carried out with skill, it not only ensures good keeping properties, but adds the hall-mark of expertise.

The object of bottling

This it can be said is to provide a method of storage which will protect the wine from the effects of oxidation and micro-biological deterioration and make available convenient quantities for consumption when required. Wines can also be further aged to provide fully matured bottle-aged wines.

The choice of bottle

Over the years, the various wine producing countries have carried out experiments to determine the effect of different coloured glass bottles on the contained wine. Results which are to be found in most textbooks of commercial winemaking are rather inconclusive. It is, however, possible to say that wine contained in dark coloured bottles usually shows less tendency to oxidise than the same wine in light coloured bottles. Having said this it is then necessary to remind the reader that white wines, containing as they do a higher dose of sulphur dioxide, are much less likely to oxidise if stored for relatively short periods in clear glass bottles than red wines. It is the usual practice—one hopes for the scientific reason given above but more likely due to tradition, coupled with long experience—to store red wine in dark brown, black or green bottles, rosé wines in clear glass, so that the eye appeal of the wine is evident and white wines in either green-tinted or clear glass bottles.

The various bottle shapes and the wines they are used to contain are given below and on page 95.

The punt in the bottom of some bottles has its origins buried deep in antiquity. One theory is that it was due to the technique of bottle making used in some places. Originally blown bottles had rounded bottoms, and would not therefore stand alone, so a base to the bottle was then formed by holding a domed former on the molten glass. This not only gave a flat solid base so that the bottle could stand upright on its own but increased the weight of glass at the bottom, giving better stability. A second school of thought says the punt was deliberately put into bottles to assist the formation of compact lees and allow easy decanting. Yet a third theory is the punt was included to give strength.

The glass moulding idea seems to have a sound historical footing although, as those who care to visit the wine museums of the world will find, it was not a universally practised technique. Assisting the formation of lees seems a little weak and as many people will know, if a slight deposit forms on the punt of a bottle it looks unsightly. The theory that the punt gives strength is borne out by engineering observations and it is because of this added strength that Champagne bottles are so shaped.

Bottle shapes in current use

Burgundy: Rounded, long shouldered bottle with a punt. Dark green or dark brown for red wines: green tinted or clear glass for white wines.

Bordeaux: Square shouldered and punted, dark green or brown for red wines, colourless for white wines. An apparently smaller "Bordeaux" bottle, with a flatter punt, has in fact the same cubic capacity, but is lighter. The styles of wine contained in this shaped bottle are considerable, from dry red wines to sweet white wines.

Hock: Tall, very sloping shouldered bottle used in Germany and France. Either mid brown or green. Usually used for white or rosé wines but sometimes light red wines are bottled in this shape. The bottles for rosé wines are made of clear glass so that the delicate colour of the wine can be properly appreciated.

Champagne: Sloping shouldered and punted. Very similar to a Burgundy bottle but a little more squat. Usually mid green but some are colourless; white or pink Champagne and other sparkling wines are bottled in this shape.

Fortified wine: The LONDON bottle is the most widely used but the distinctive round-shouldered BRISTOL bottle is used by a large firm of wine shippers and several smaller wine merchants in the City of Bristol. Usually dark brown or dark green bottles.

Lond. bottle

Bristol bottle

All the above bottles, with the exception of the London and Bristol bottles, are sealed in commercial practice with drive-in corks.

The Champagne bottle is sealed with a cork made from a cork laminate and is wired on to the bottle neck.

Both the London and Bristol bottles have flat or domed top stopper corks. These corks are sometimes referred to as flanged.

Cleaning and sterilising bottles

All bottles should be thoroughly cleaned with hot water and a detergent, using a good bottle brush to reach all the difficult angles and finally rinsed out with warm, fresh water. Bottles should be allowed to drain in an inverted position, thus allowing the maximum removal of water and the minimum entry of airborne impurities.

To sterilise the bottles, fill each with a 1/80 solution of Milton or a 2% sulphur dioxide solution (dissolve 5 ozs. or 150 gms. of sodium metabisulphite powder in 1 gallon of water). Allow to stand for 15 minutes then drain out the solution, retaining it for further bottles. Finally, swill each bottle out with a little *boiled* tap water. (The reason for this is given in chapter 1). The bottles are now ready for filling and should be left inverted until required. Alternatively, after cleaning, bottles can be sterilised in a pressure cooker if it is of sufficient size.

Choice of seal

With the advent of plastics an alternative to cork now exists but the merits of both need careful consideration before deciding which to use. Certainly plastic stoppers are much easier to clean and use again but they do not give the same tight seal as cork and this is a most important detail if bottles are to be stored in a horizontal position.

Resealing with a plastic stopper is easier than with a cork but if half and full bottle sizes are filled at the time of bottling, this is often unnecessary as a half bottle is usually consumed easily, and does not require resealing.

If the considerations above are of no significance, the choice boils down to one of individual preference and cost.

As corks are very much cheaper they are chosen by many amateur winemakers but some experimentalists are using plastic.

Straight sided corks can be purchased in two lengths, one about (3.2–3.8 cms.) and the other (4.4–5.0 cms.). Either will do very well but if it is intended to store a wine for a long period of time, it is advisable to opt for the longer of the two as this will provide less risk of wine seepage.

Corks selected for wine use should be smooth and have a regular grain with close veining. Fig. 18 shows what is meant.

 Fig. 18

The cork on the left of the diagram is of high quality whilst those extending to the right are of progressively lower quality.

It is important that the corks should have a certain spongy feel and not be hard. The spongy feel is given by the air contained in the pores of the cork and allows more effective sealing, hard corks showing a greater tendency to shrinking. Because of the large air spaces corks are also likely to contain micro-organisms and these could, in their turn, produce wine spoilage. Consequently corks should be sterilised before use.

Sterilisation of corks

At one time it was common practice to boil corks to sterilise them but this makes them hard and brittle. Additionally corks so treated are difficult to remove with a cork screw and frequently break up, having to be dug out piece by piece.

The easiest way to sterilise corks is to soak them for two hours in 1% sulphite solution with a little glycerine added to it. The glycerine acts as a lubricant and helps to prevent hardening of the cork.

It is economic to take the 2% sulphite solution used for bottle sterilisation and dilute it with an equal volume of water, add one teaspoonful of glycerine for each gallon of solution and use this for sterilising the corks.

Two hours is quite sufficient contact time; longer periods only serve to harden the corks and allow absorption of an excessive amount of solution which will squeeze out when the cork is compressed prior to insertion into the bottle. When sterilisation is complete the excess sulphite solution should be removed by either vigorous shaking or centrifuging in the spin dryer of a washing machine for $\frac{1}{2}$–1 minute.

The spinner should, of course, be clean before the corks are placed into it and should be washed out with tap water after removal of the corks to ensure that no traces of the acidic sulphite solution remain to cause corrosion of the metal parts.

The corks may be kept sterile after removal from the spin dryer by placing them into a clean polythene bag which has been rinsed out with 2% sulphite solution, the neck of the bag being sealed with an elastic band and the corks not handled until required. Corks can be sterilised in a pressure cooker and for most practical purposes the amateur will find this method sound but the corks sterilised in this manner are not so easily worked as with the sulphite method.

Filling the bottles

This task is probably the second most pleasurable task the amateur winemaker undertakes, the first of course being savouring and drinking the results of several months labour.

The easiest method of filling a bottle is to place a clean, sterile funnel in the top of the bottle and pour wine in from the storage vessel until (4 cms.) above the shoulder of the bottle is reached. This method will, however, lead to wine entering the bottle becoming exposed to air by cascading down to the bottom or beverage level during filling.

Oxidative changes will thus be speeded up and, in the case of white or rosé wines in particular, this is most undesirable. In the same manner air sucked back through the wine in the storage vessel will dissolve and produce oxidation changes.

Some ingenious bottling methods for amateurs have been put forward. A method utilising the gases liberated from an active fermentation to replace the air contained in the bottles is particularly interesting because it is a simplification of a technique used in the bottling industry whereby air in the bottles is replaced by pure sterile filtered nitrogen or carbon dioxide. A drawback does exist in that some fermentation gases smell rather strongly and since alcoholic beverages show a great tendency to pick up "off" flavours from the atmosphere, some care must be exercised in its use.

An alternative and just as simple a method can be adopted from the technique used in many Continental wine cellars to avoid excessive oxidation. The principle of the Continental method is shown in fig. 19. The entering air or nitrogen is sterile filtered by bubbling it through 1% sodium metabisulphite solution in a specially designed trap unit.

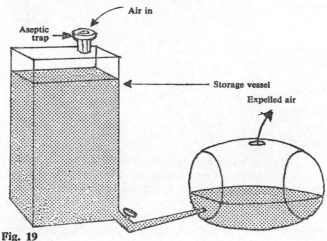

Air in

Aseptic
trap

Storage vessel

Expelled air

Fig. 19

It is possible to utilise the same principle for bottling amateur wines. First the storage vessel is fitted with a syphon and vent tube (the method for operating this has already been described in chapter 6). The syphon tube needs to be flexible for most of its length so that the control clip can be applied high up and the tube extended to the bottom of the bottle. Obviously the syphon tube must be clean and sterile if trouble is to be avoided.

After the syphon has been made and the clip applied a trap bottle of the type referred to in chapter 7, but containing a 1% solution of sodium metabisulphite is connected to the vent tube.

When the clip on the syphon tube is opened wine flows and air is sucked through the sulphite solution to replace it. In passing through the solution it is sterilised and cannot introduce any spoilage organisms into the wine. The assembly in use is shown in fig. 20.

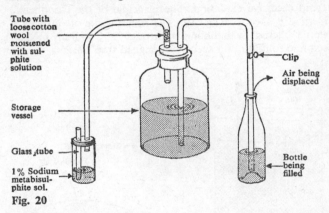

Fig. 20

After the bottle has been filled to a level that will leave only a very small air space when the cork has been inserted, the flow of wine is stopped, the bottle removed and the next one filled.

Do not leave filled bottles for more than a few minutes before inserting the corks; in this way possible contamination will be reduced to a minimum.

Sealing the bottle

An important point to remember is that when a cork is pushed into the neck of a bottle, the air above the wine is compressed and will tend to resist entry of the cork. A simple but primitive way of overcoming this difficulty is to insert a stainless steel wire into the neck of the bottle, put the cork on to the bottle and drive it in with a "flogger." This is not a sharp salesman but a piece of heavy flat wood with a handle.

When the c ork is nearly in the wire is withdrawn, thus allowing any air to escape in the channel made by it. A piece of thin string is also effective in this way. The cork is then driven fully home. This technique is still used but where the wire or string has been withdrawn a channel often exists which will allow wine to seep when the bottles are stored in a horizontal position.

A more satisfactory method of corking is to use a hand corker constructed of metal. Devices made from wood are much less satisfactory because when a bottle is sealed a little wine often spurts out. This will soak into wood and can become infected with spoilage organisms which in turn will find their way on to corks subsequently applied in the apparatus. The metal tool is of course impervious and can easily be washed clean. An inexpensive metal corking tool is shown in fig. 21.

The cork is placed into the tool in the open position. It is then closed and the two handles gripped together, compressing the cork. The corker is then placed over the neck of the bottle and the lever operated to drive the cork

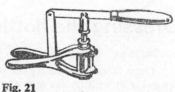

Fig. 21

home. It is, of course, helpful if a second person holds the bottle during the latter operation covering his hands with a thick cloth or stout gloves. Alternatively, a bottle size hole in a large chunk of wood will hold the bottle steady during the corking operation.

Avoid the use of a copper or bronze corker; copper contamination of the wine is then easily avoided from this source.

Champagne corks used for sparkling wines have to be wired down to prevent them being blown out again; the technique is shown below:

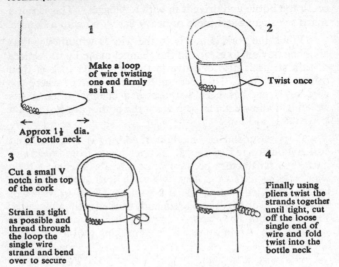

1

Make a loop of wire twisting one end firmly as in 1

← Approx 1½ dia. → of bottle neck

2

Twist once

3

Cut a small V notch in the top of the cork

Strain as tight as possible and thread through the loop the single wire strand and bend over to secure

4

Finally using pliers twist the strands together until tight, cut off the loose single end of wire and fold twist into the bottle neck

Dressing the bottle

Sealing wax

Once it was customary to seal the tops of table wine bottles with sealing wax and impress into this a seal. Nowadays this practice has almost ceased even in commerce except for vintage port. No doubt the mess and inconvenience of removing the wax has played a large part here but it cannot be denied that this old method gave good protection to the top of the cork. All that is required is a small tin of sealing wax kept molten over a low flame into which is dipped the dried top of the cork and bottle neck. When withdrawn the bottle is kept inverted for a few seconds to allow the wax to set partially, and then turned up the right way and a seal ring or other appropriate object pressed into it.

Lead and alloy capsules

These are used quite widely commercially. The capsule is crimped on to the bottle by a special machine but the amateur can do quite a competent job by hand-rolling. Many people have misgivings about lead capsules as any wine which seeps through or past the cork when the bottle is horizontally stored will, due to its acidic nature, attack the lead and form a lead salt crust at the top of the cork. Providing the bottle is cleaned off properly before the cork is withdrawn, this will not present a health hazard but lead salts are poisonous and care should be taken over cleaning.

For those who do not wish to use lead capsules alloy foil ones provide an alternative and are in every way as attractive in appearance.

Shrink fitting capsules and sleeves

In the past ten years this type of final seal has found a wide commercial market and many amateur winemakers use them. They are made of viscose and stored in a solution of glycerine and water containing preservative substance in which they remain flaccid and pliable. When fitted over the neck of the bottle the viscose is quite slack but in a warm atmosphere as water is lost it will tighten up and assume the contour of the bottle.

The seal produced is attractive in appearance especially if chosen to match the colour of the lettering on the label.

Two disadvantages exist, however.

1. If the bottles are stored in a moderately warm and dry atmosphere the viscose material tends to lose too much moisture and becomes brittle, often splitting the bottom or top of the seal.

2. Should a cellar be used for bottle storage, where the atmosphere is slightly damp, the viscose will take up water and become loose again. Whilst this can be reversed by placing the bottles in a warm spot, it defeats the intention of capsuling which is to provide a seal for the top of the cork and protect it from insect or fungal attack at the same time presenting an attractive finish.

Capsules are now on the market which do not have to be stored in a liquid. These have to be stretched over the neck of the bottle to effect a seal. Sometimes air is trapped under the capsule and it will not sit down properly but this can be overcome by making a pinhole in the top to allow the air to escape.

Labelling

The choice of label design is, of course, one of personal preference. A wine circle may find one of its members has artistic talent and can design a distinctive and individual label which can be printed locally and bought by circle members.

Alternatively, where individuals are concerned, it may be cheaper to buy the standard label designs which are available from "The Amateur Winemaker", South Street, Andover; there is one in the range to suit anyone's taste.

Whichever course is chosen, an effort should be made to balance the label and capsule colour as eye appeal is important. In commerce very many products are sold on the basis of this consideration.

Do's and Don'ts

DO remember bottle and cork sterilisation is important.

DO use straight sided corks when wines are to be stored horizontally.

DO purchase good quality corks—they are cheapest in the long run.

Do take care when filling bottles and try not to include more air than necessary.

DO seal the top of corks to prevent insect attack.

DO use a metal corking tool—but not a bronze one.

DO label bottles attractively: it adds distinction.

DO NOT believe bottling "technique" is unimportant: it matters a great deal.

DO NOT forget bottle shape and colour influences presentation.

DO NOT use stopper or flanged corks if bottles are to be stored horizontally.

DO NOT use a wooden corking tool they absorb wine and can be troublesome.

DO NOT think bottle dressing is unimportant or "gilding the lily": cleverly done it greatly enhances the beverage when eventually served.

CHAPTER 10

Storing and Serving

NOW the wine is safely in the bottle, attractively labelled and capsuled, the next consideration is where and how to store it until the great moment arrives when it is required for drinking. And when this great moment arrives, the actual serving of the "ambrosia" to which so much attention has been given, must also receive due consideration. It may be necessary to decant, chill or slightly warm the wine before serving and the choice of glass size and shape is important to the eye as well as to the senses in order to appreciate to the full the quality of the wine.

Storage

What are the broad requirements for a cellar? These can be summarised as follows:

1. Good ventilation to prevent mustiness and excessive condensation.

2. Absence of light. As already explained, ultra-violet light greatly hastens aging but wines stored in the dark mature gradually in a more predictable manner.

3. A reasonably stable, cool temperature 52–58°F. (11–14°C.) summer and winter is ideal. It is sometimes difficult to maintain but 70°F. (21°C.) should not be exceeded unless a rapid acceleration in aging is acceptable and expected. A maximum and minimum thermometer in the cellar provides an easy means of checking the temperature fluctuations and if a note book is kept into which observations are recorded, it is possible to see exactly the variations during the seasons of the year and whether any attention is required to thermal insulation or heating.

Where is the best place to store the wine? In days gone by most houses had underground or semi-basement cellars and for winemakers who are fortunate enough to have such cellars nowadays, wine storage can be planned in a grand manner. However, as already mentioned, it is essential for any cellar to be well-ventilated and this point should particularly be borne in mind with underground storage otherwise problems of condensation and mould growth will soon arise. These are far from funny when all the effort put into labelling is ruined. An additional advantage with the subterranean cellar is that the temperature is fairly constant both in summer and winter.

Fortunately for the majority of winemakers it is unnecessary for a cellar to be underground but the aboveground counterpart needs a little more attention and care. A cupboard under the stairs may provide the solution, usually being dark and cool if it is away from a chimney or central heating system.

A disused box-room can be turned into an excellent cellar with the windows fitted with a blind or screen to keep out the light—especially sunlight—to avoid an acceleration in aging with unpredictable results.

A garden shed or portion of the garage might also well be turned into a cellar providing attention is given to likely temperature changes and with requirements of thermal insulation or heating.

Having decided where the cellar is to be situated, the next consideration is how to store the bottles.

Positioning the bottles

All bottles corked with short drive-in corks or stopper (flanged) corks should be kept in a vertical position. This is because the length of cork in contact with the glass of the bottle is invariably insufficient to prevent wine seepage over a period of time if the bottles are stored horizontally.

Fortified wines also should be stored in a vertical position because suberine, a major constituent of cork, is soluble in alcohol and with the continual contact during horizontal storage, the possibility of this dissolving is greatly increased with the consequent likelihood of seepage.

Table wines corked with drive-in corks and sparkling wines which have a special type of stoppered cork, should be stored horizontally. Storage in this position keeps the cork moist which tightens it and prevents seepage occurring. In the case of sparkling wines, horizontal storage provides a better seal and prevents loss in the gas content which gives the sparkle.

It must be remembered that if a cork is allowed to dry out, not only does a danger of seepage exist but also the problem of air entry. If this occurs, acetification starts with some of the wine alcohol changing to vinegar.

Turning for a moment to commercial practice—it should be explained that vintage port is corked with drive-in corks of a similar type to those used for table wines and the top of the cork is sealed with either sealing wax or a special plastic cap. The bottles have a stroke of white paint on the side which is uppermost in the storage rack and they are stored on their sides to mature for a period of maybe 20 years or more. How then does this fit in with what has been said above about fortified wines and suberine solubility?

The sealing wax provides an additional seal and this prevents seepage but, until a few years ago, this was the only protection and when the seal was removed often some evidence of seepage out as far as the seal could be seen. Now, however, corks are being produced which have been treated with wax in such a way as to render the cork much less open to the entry of wine when the bottles are stored horizontally and these corks have found wide favour both for table wines and vintage port. The methods of storage can now be studied.

Binning

Bottles can be stored in a variety of ways, depending on the number of bottles to be accommodated and the sum of money which can be expended. If a lot of bottles of the same wine have to be stored for a relatively long period of time, binning is the cheapest method.

In its simplest form, all that is required is a flat floor and two walls or partitions between which the bottles can be stacked. A row of bottles is placed, necks facing out, at the rear of the compartment, then a second row positioned with necks facing in, to coincide with the gaps between the necks of the first row. This process may be repeated as many times as the ground area allows and then continued into a second and subsequent rows. The principle is shown in fig. 23.

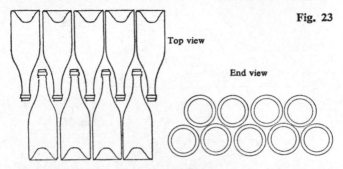

Fig. 23

Top view

End view

A more sophisticated system of binning involves the use of supporting laths and gives greater stability to the stack of bottles and the floor does not have to be level.

A thin lath is placed upon the floor and shims inserted under it at intervals until it is level, as shown by a spirit level. A bottle is then laid upon the lath so that its base is projecting about one inch beyond the lath edge. A second lath is then taken of sufficient thickness to support the neck of the bottle and level it. This can also be done with the aid of a spirit level and the insertion of shims if necessary. The whole length of the lath can now be filled with bottles. When the next row is to be built, a lath sufficiently thick to hold a bottle at the neck height of the first row is required. This is placed on the floor and levelled as before. The second row of bottles is then positioned. fig. 24 shows how the bottles and laths are placed

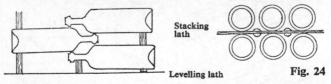

Stacking lath

Levelling lath

Fig. 24

in the first two rows. The two systems described above are about the most compact ways of positioning Bordeaux bottles but it is possible to bin Burgundy and Champagne bottles, because of their sloping shoulders, in a slightly different manner and the method is shown in figs. 25 and 26.

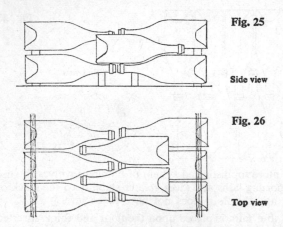

Fig. 25

Side view

Fig. 26

Top view

When a variety of wines in smaller numbers have to be accommodated, the pigeon hole rack or wire frame type can prove most useful. Both types can be put side by side with similar units to form a relatively large cellar. Figs. 27 and 28

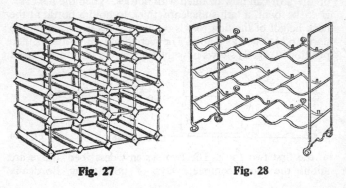

Fig. 27

Fig. 28

show both these patterns. And lastly, for those who want the cheapest, if not the most convenient storage container, the divided packing cases used by wine merchants to transport table and fortified wines will, if the top flaps are cut off and the cases turned upon their sides, make good storage containers.

Decanting

During the time spent in bottle, it is common for red wines to throw a sediment or deposit. This, as explained earlier, is due to the oxidation of tannins, polyphenols and other substances associated with colour over the period of storage. Whilst

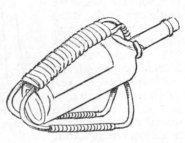

Fig. 29

not harmful, deposits of this nature are unsightly and the usual practice is to remove the clear wine from the sediment before serving. This procedure is called decanting.

This separation of clear wine from the sediment which has formed in the bottle during storage requires a little care and practice if the minimum of wine is to be wasted.

Anticipation is rather important; if the required wine is removed from the horizontal storage position and stood vertically for 24 hours, the lees settled along one side of the bottle will drop to the bottom of the bottle and make decanting an easy operation. If this is not possible, the bottle should be removed from storage very carefully, taking care to keep it in the horizontal position and not shake or turn it, and then be placed in a serving cradle (fig. 29 shows one of these) where it should be left for an hour or so in case the sediment has been slightly disturbed.

111

Having either allowed the bottle to stand for a day or placed it in a cradle, the next thing to do is to remove the capsule or sealing wax protecting the cork. This should be done with a sharp pen-knife and care taken not to shake or disturb the bottle. After removing the seal or capsule, wipe the neck of the bottle clean with a damp cloth or tissue. The cork can now be removed. This should on no account be done with an old-fashioned cork screw which requires a straight pull on the cork as the deposit will undoubtedly be shaken up and a great deal more wine wasted than necessary. Choose a cork screw with either

Fig. 30

a leverage or self-pulling action. In this way the bottle will not be shaken. Fig. 30 shows an example of each type. Having removed the cork, wipe the inside of the bottle neck with either a clean cloth or tissue.

The actual decanting can now be performed. A clean decanter or empty bottle is first placed upon the table and a lighted candle or small table lamp placed next to it so that when the wine is transferred, the light will fall on the shoulder of the bottle from which the wine is being poured. The bottle from which the wine is to be decanted is held steady in the hand and the wine carefully poured using the light to indicate when the sediment just starts to come over, at this point decanting is halted. If this operation is carried out carefully, only a little wine will be wasted. It is not uncommon to see the use of filter paper or cloth advocated for decanting but the use of these will not lead to the high quality results obtained by the method described above, for the reasons explained

112

in the chapter on fining and filtering. Decanting with a filter is the same as filtering a wine, using a folded filter paper, which process has already been described.

Serving the wine

As a general rule, red wines are served a *r*room temperature but white, rosé and sparkling wines are served chilled. This is only intended as a guide but there are sound reasons for paying it attention.

One only has to experiment by drinking a cold red wine and drinking the same wine at room temperature, say 68–70°F. (20–21°C.), to discover the delight of a more pronounced "nose" as well as a decided improvement on the palate. By the same token, warm white wine is not pleasant to taste and loses a considerable amount of its charm.

On the other hand, white wines improve when slightly chilled, being "crisper" than at room temperature but care should be taken not to over-chill because the substances which contribute to the bouquet are much less volatile at low temperatures and thus this important feature of a wine can become impaired.

Aperitifs are usually found to be at their best when served slightly chilled but dessert or fortified wines may be served at temperatures depending upon circumstances and personal choice.

The choice of glass

A lot has been written about the right type of glass to use for different styles of wine and this is where, to a certain extent, art takes precedence over science, but both factors have to receive consideration.

The initial, almost sub-conscious, appraisal of a wine is the way in which it is served—the colour of the wine, the salver and glasses all contributing to the general sense of pleasure. It is obvious that a brightly ornamented salver will spoil the general appearance of the offering and similarly coloured or decorated glasses will detract from the appreciation of the wine's colour. For this reason, clear glass is superior but who can deny the enjoyment of taking a cut glass of wine from a silver salver? The cut crystal, whilst having a pattern which perhaps prevents a clear sight of the wine, at the same time adds an elegance which enhances the moment of appraisal.

The best shape of glass to suit the wine being served has to be considered. Red wines benefit from being served in a relatively short stemmed, bowl shaped glass because such a glass will fit into the hand easily and the heat of the hand will bring out the bouquet of the wine.

White, rosé and sparkling wines need to be served in long stemmed glasses with a smaller bowl than that of a red wine glass. In this way the natural tendency is to hold the stem and not the bowl of the glass, thus keeping it cooler and the smaller bowl retaining the bouquet of the wine better as it is not so open to the atmosphere. Shallow saucer glasses are not very satisfactory for sparkling wines as they allow the sparkle to escape easily, a tulip shaped glass being more suitable and, furthermore, doing double duty as an excellent white or rosé wine glass.

Fortified wines are served in smaller glasses for the very good reason that they contain a higher amount of alcohol and if served in the quantity usual for table wines, would soon render most of the company incapable!

Tasting glasses are tulip-shaped but with shorter stems than the glasses described above. This gives good stability and provides for excellent bouquet retention. Whichever glass is used, it is essential that it should not be filled to the top because, in this way, no space would be left for the bouquet, which is so important. Two-thirds of the glass filled with wine is a good guide except for tasting of a serious nature, when half-full should be the rule. Fig. 31 illustrates each of the glasses described above.

To conclude this chapter a few suggestions are made when to serve the various wine styles and the foods they complement. However, these observations are meant to serve as a guide only, as it should be remembered that wine drinking is an object of personal pleasure, no two palates are alike and, whilst experience will guide the discerning imbiber as to the wines he prefers, rigid adherence to a certain style of wine purely because "it is the custom" should not be encouraged.

Fig. 31

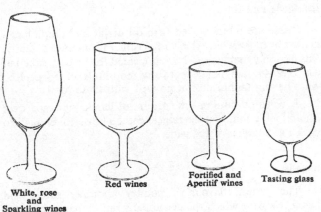

White, rose
and
Sparkling wines

Red wines

Fortified and
Aperitif wines

Tasting glass

Wines and food

Aperitifs

These are designed primarily to stimulate the appetite, to produce a feeling of relaxation and anticipation before a meal. It is generally agreed that dry or slightly sweet wines, rather than decidedly sweet wines, have this effect on the majority of people accustomed to drinking but it should not be assumed that drinking a sweet wine before a meal is wrong. The range of aperitifs is very wide and very much a case of personal preference.

Spicy foods

Or those with highly seasoned dressings have such powerful flavour attributes that any wines with fine or delicate flavours will be overpowered and much of the wine's appeal will be lost, For these foods use an "everyday" wine, one that has some flavour but is not of sufficient character to warrant keeping for a more auspicious occasion.

Sea foods and fish

These are often cooked with oil or fat and are complemented by a dry white wine with a rather pronounced acidity. This not only manages to be appreciated for its own sake but also cleanses the palate between mouthfuls of food. Red wines having tannin do not go well with these foods.

A word of caution—the aroma of lemon squeezed over fish and left behind on the fingers can be powerful enough to mask the bouquet of the wine.

Poultry

The natural flavour of poultry is rather subtle and therefore a wine which will accompany rather than overpower it needs to be chosen. Either a dry white, rosé or red wine will do admirably, providing there is not a lot of tannin, excess acidity or over-sufficiency of fruit flavour.

Meat and game

These dishes have a pronounced robust flavour and a full-flavoured dry red wine is ideal with the red-blooded meat with its luscious juices—the tannin of the wine becomes smoothed off and the flavour of the meat is enhanced. A good vintage dry red wine such as an elderberry or pomegranate shows off to advantage with a roast—each mouthful of meat demanding more wine and every drop of wine cleansing the palate and calling for more food!

Sweets

The food, as its name implies, contains a relatively large amount of sugar and a sweet wine is most suitable.

116

Dessert, cheese or coffee

Excellent with a fortified dessert wine.

Sparkling wines should not be forgotten and these will satisfy the requirement of any course, or act as an aperitif. The effervescence of the wine provides zest which also adds to the success of the meal.

Do's and Don'ts

DO make sure that your cellar is well ventilated and dark.

DO try to keep an even storage temperature.

DO store wines with drive in corks in the horizontal position.

DO choose a storage method which is best suited to requirement.

DO pick a corkscrew that does NOT require muscles.

DO chill white, rosé and sparkling wines before serving.

DO pick a glass which is suited to the purpose and to the wine.

DO NOT store bottles in a damp or wet cellar.

DO NOT allow sunlight to fall on bottles being stored.

DO NOT let bottles become excessively cold.

DO NOT store fortified dessert wines in the horizontal position.

DO NOT heat red wines before serving.

DO NOT be dogmatic about what wine to drink with a particular food.

CHAPTER 11

How to evolve recipes

MOST books on winemaking give a comprehensive selection of recipes but this book has been written to encourage the beginner to learn the basic principles of winemaking, to carry out the various procedures in the correct order and to foster the desire for beginners and experienced winemakers alike to experiment in order to build up knowledge in the amateur field of winemaking.

This chapter, then, does not set out recipes to be slavishly followed but is written in an effort to help and encourage the amateur winemaker to plan and evolve his or her own recipes.

Let it be said at the outset that this needs care and thought and detailed notes must be kept in order to gather information and to adjust, amend, vary or repeat succeeding fermentations. It is true these precise and careful preparations are more painstaking than mixing given ingredients together and "hoping for the best" but the advantages of making balanced wines of the styles desired more than compensates for the extra care involved in the initial stages and after a while the procedures are followed naturally. When labour, time and money are to be expended on the various stages of winemaking, who can deny that a little extra care and calculation at the beginning is most worth while? If records are kept, together with tasting notes, useful knowledge and experience will be assimilated and a background to quality winemaking achieved.

As in so many aspects of this hobby, one can spend as little or as much as one likes on ingredients. People have been known to try to make wine from cold rice pudding, mashed potato or from the water left from boiling Christmas puddings, and although these ideas are not advocated in this book, they go to show how great is the scope for the amateur to experiment with the variations and combinations of ingredients available.

With all ingredients it is most important that they are as fresh as possible and in peak condition. Under-ripe fruit only produces an unpleasant flavoured beverage whilst over-ripe fruit contains a high micro-organism population which needs careful sterilisation and in any case the resultant wine is of poor flavour and quality. Think of eating a mushy over-ripe pear and the inference is obvious. Do not be carried away with enthusiasm if the greengrocer offers over-ripe fruit at a "knock-down" price. The task of cutting away all the mouldy pieces and being left with a small quantity of useable fruit makes it apparent that it is more economical to buy sound fruit at a higher price and the task of preparation is much more pleasant, with a good end-product more assured.

True, one hears occasionally of winning wines made, say, from mouldy oranges, but it is a certainty that the maker has been lucky and the practice of using unsound fruit is a gamble, with the odds against the maker.

Commercially, of course, grapes are the main ingredient for wine and one often hears the phrase "the grape has everything" but for the amateur, using grapes does not always ensure a superb wine. It has to be remembered that the grapes used commercially are specially grown and chosen against a backcloth of years of experience of climatic and soil conditions, together with a vast knowledge of the varieties which are suitable. Grapes for home winemaking can be bought from the shops—but often the wrong varieties are available to produce a good wine—and grapes grown in gardens in the United Kingdom battle against difficult climatic conditions with the lack of sunshine producing small, acid, under-ripe fruit usually unsuitable for winemaking. More and more people are planting vines for winemaking purposes but the above points have to be borne in mind.

However, the amateur winemaker has a wealth of different fruits from which to pick, some producing a more balanced wine than others and it is often necessary to use more than one variety to achieve a balanced effect without recourse to blending finished wines.

Beginners to the art of winemaking and wine tasting are often vexed to know what a balanced wine is. If one's palate needs educating in this respect, commercial wine tasting, accompanied by the tasting notes of an expert, is a fine habit in order to set standards and gain knowledge and experience of the texture, flavour, sweetness, acidity and alcoholic content of wines as well as the ability of choosing the right wine to go with various foods. An amateur cannot emulate, for example, a vintage Burgundy but an educated palate can appreciate and savour commercial and home-made wines alike for their individual attributes. Some winemakers have formed groups to taste wines with food and have found the resultant experience very useful. Commercial and home-made wines can be tested beforehand by the host or hostess for specific gravity and acidity and then, during the tastings, personal preferences and evaluations can be made as to the acceptability of the acidity, sweetness or dryness, balance, etc., of the wines with the dish served, notes as to "value for money" where relevant, together with estimated gravity and acidity readings. It can be surprising how comments vary and how wrong one can be when comparing notes with the actual readings.

The question the beginner asks is how to begin wine-making but without a recipe as a guide? This is not as difficult as it might appear. Commonsense and thought are the main ingredients for preparation together with reference to the previous chapters in this book, in particular chapter 3, for the procedures of juice preparation and calculations of sugar and acid corrections.

Once the choice of fruit to be used has been settled, the next step is to decide the style of wine required, as this is important not only in regard to the sugar and acid corrections which will have to be made, but in connection with the amount of flavour and body to be produced in the wine.

For the beginner, it is as well to start with wines light in alcohol and acidity and on the sweet side, as these will not only be more acceptable to the palate but will also not need as long maturation as the heavier bodied, more alcoholic, full-flavoured styles—a point to be borne in mind when one is longing to taste the first brew!

Mention has been made of various wine styles in chapters 5 and 10 but these are given again in the following table to show the generally accepted styles of wines to accompany food.

As will be seen from the table overleaf, "aperitif" wines cover a vast range: they can be sweet, dry or medium flavoured to suit one's palate. Commercially the range is extensive and, as the essence of an aperitif is to cleanse the palate and stimulate the appetite, generally a dry, relatively high-alcohol wine with a piquant flavour fits this role, the fino sherry being an example in the commercial world but, in this case, as with any choice of wine, preference is a matter of individual palate.

The term "social" wine is decried by some but in amateur winemaking spheres it has come to mean a wine which is drunk with savouries, a little stronger in alcohol, of medium acidity and slightly sweeter than a table wine and one which has found acceptance at Circle meetings and when friends call. Of course, it can also be argued that *all* wines and spirits are "social."

"Table" wines are self-explanatory and are made to accompany the various courses of a meal, cleanse the palate and stimulate the appetite. Here the tasting groups mentioned previously are particularly useful in deciding the variety of wine to accompany various dishes.

"Dessert" wines should be fully flavoured, with high alcohol and heavy body and these wines take longest to mature. A higher quantity of fruit to the gallon is used and for some time after fermentation the wine will taste over-fruity. To obtain a true dessert style wine it is necessary to fortify (see chapter 6) as home-made wines will not attain a higher alcohol content than approximately 15% v/v. The fortification will balance the sweetness necessary in this style of wine, stabilise it and give good keeping properties—so for a few extra shillings, one can achieve a truly professional quality.

121

TABLE 4

Food	Wine colour	Style	Approx. S.G.	Approx. Alcohol %	Flavour	Approx.% Acidity as citric	Bouquet	Body
Nuts, crisps, etc.	White, red, rosé, brown	Aperitif	0.990/1.010	15—20	Light, medium, full	0.50-0.75	Medium, full	Light, medium, full
Savouries, sandwiches	White, red, rosé	Social Med. dry	0.998/1.000	14	Light/medium	0.65-0.75	Medium	Medium
Spicey	White, red	Table Dry. Vin ordinaire	0.990	12	Medium	0.7	Light	Light/medium
Sea food and fish	White	Table dry	0.990	11—12	Light	0.70-0.85	Light/medium	Light
Poultry	White, red rosé	Table dry Med. dry	0.990 0.998	12 12	Light	0.57-0.70 0.75	Light/medium	Light
Meat and Game	Red	Table dry	0.990	12—14	Full	0.57	Full	Medium
Sweets	Red, white	Table dessert Sweet	1.020	14	Medium	0.70	Full	Medium
Dessert	Red, white brown	Dessert Sweet	1.038 1.038 1.038	18—20 18—20 18—20	Full Full Full	0.45 0.50 0.50	Full Full Full	Full Full Full

Alternatively, additional sugar can be added at the onset of fermentation to leave residual sweetness after fermentation but this is sometimes an uncertain operation, due to sticking of an over-sweetened "must."

To obtain the body and fullness necessary with a dessert wine, without being over-flavoured and over-acid, thought must be given to blending ingredients and dried fruits are excellent for this purpose, raisins and sultanas being particularly useful. Dates, figs and prunes can also be used to advantage but care has to be taken not to add too many due to their high flavour, which can obtrude. Rose hips and rose hip shells add body to a wine, as do fresh bananas, and notes will be found relating to these later in the chapter. Honey and grape concentrates are also useful additions to a dessert wine, the honey giving a rounded finish to the flavour.

One soon learns that elderberries make an excellent dessert style wine which, having a strong flavour of their own, can be balanced with the addition of other ingredients, as suggested above, to make a wine with the necessary balance of body and flavour. Examples are given later in the chapter, under the heading "elderberries" in the recipe section.

Having decided upon the style of wine to be made and the main fruit to be used, the next and most important step is preparing the "must." The quantities and blending of additional ingredients have to be considered, as referred to above under dessert wines.

The first consideration is flavour. Naturally, if a light flavoured wine is required, a smaller quantity of the main ingredient will be used than if a full flavoured style is desired. Also, although some wine styles call for a full flavour, this does not mean over-flavoured. No wine should be so over-flavoured that all other attributes are lost, making the beverage taste like a cordial rather than a wine. Conversely, no recipe should contain such minimal quantities of ingredients as to produce a characterless, unbalanced drink with no wine-like qualities. Do not try to be too economical with ingredients, this results in very little goodness and nutrients for the enzymes in the yeast to work upon and produces a ferment which "plods on" for months and an end-product which is not worth drinking.

If only a small quantity of the main ingredient is required to produce the flavour required, then add other low flavoured ingredients to provide the body and constituents to make a balanced wine. Bananas, rose hip shells, sultanas, raisins, concentrates, honey or any low-flavoured juices can be added to give quality to the wine.

This adjustment of quantities and blending of additional ingredients at the initial preparation stage becomes "second nature" after the first few wines have been made and will, no doubt, be apparent from a study of the recipes at the end of this chapter.

It is possible, of course, to blend wines after fermentation but additional time is often involved as invariably blending results in secondary fermentation—whereas careful adjustment of ingredients at the "must" stage makes final blending unnecessary.

A table is given on the next page showing popular ingredients available to the winemaker and their suitability for differing wine styles, followed by general details relating to the ingredients and their preparation, and some recipes for guidance.

Information given in the table can only be general, as it will be appreciated varieties of the same fruits can differ in composition, as can the sweetness and acidity from year to year, depending upon the climatic conditions during which the fruit ripened. However, it is hoped that the details shown will enable the reader to decide which ingredients to use for the wine style desired. The necessary acidity and specific gravity adjustments can then, of course, be made in accordance with the tables and tests given elsewhere in this book.

Finally, during preparation of the "must," it is always a good idea to taste the juice critically to build up knowledge of flavour, acidity, tannin and sweetness from the very inception of the wine. The palate and nose are very sensitive organs and indeed, some people with years of experience in winemaking, have enough palate sensitivity to rely upon their taste buds to tell them whether the "must" is balanced or otherwise. The acceptability of the flavour, smell and acidity of the juice to the palate and nose at the preparation stage is a very good guide to the resultant wine.

TABLE NO. 5

GENERAL DEGREE OF ACID, FLAVOUR, TANNIN AND SWEETNESS TO BE FOUND IN VARIOUS POPULAR WINEMAKING INGREDIENTS AND THEIR SUITABILITY FOR DIFFERING WINE STYLES

1. Fruits which produce balanced wines used as the main ingredient

Fresh:	Tannin (mainly in skins)	Acid	Sugar	Pectin	Flavour	Body	Wine style suitability
Apples	Low/med	Med/high	Med.	Low/med	Low/med	Low	Table/Med. sweet
Apricots	High	Med.	Med.	Med.	Med.	Med.	Table/Med. sweet/Dessert
Bilberries	High	Med.	Med.	Med.	High	Med.	Table/Med. sweet/Dessert
Blackberries	Med.	High	Med.	High	Med.	Med.	Table/Med. sweet/Dessert
Blackcurrants	High	High	Med.	Med.	High	Med.	Table/Med. sweet/Dessert
Cherries, sweet	Med.	Med.	Med.	Med.	High	Med.	Table/Med. sweet/Dessert
Cherries, cooking	Med.	High	Low	Med.	High	Med.	Table/Med. sweet/Dessert
Damsons	High	High	Med.	High	High	Med.	Table/Med. sweet/Dessert
Elderberries	High	Low	Low	Med.	Med.	Low	Table/Med. sweet
Gooseberries	Low	High	High	Med.	Med.	Med.	All wine styles
Grapes	Med.	Med.	High	High	High	Med.	Aperitif/Table
Grapefruit	High	High	Low	High	Med.	Med.	Table Dessert/Dessert and Med. sweet
Loganberries	Med.	Med.	Med.	High	High	Med.	
Mandarins	Med.	High	Med.	Med.	Med.	Med.	All wine styles
Oranges, sweet	High	High	High	High	High	Med.	All wine styles
Oranges, Seville	High	High	Med.	High	High	Med.	Aperitif
Peaches	High	High	High	Med.	High	Med.	All wine styles
Plums	High	High	High	High	Med.	Med.	All wine styles
Pomegranates	High	Med.	High	Low	Med.	Low	Table/Med. sweet
Raspberries	Med.	High	Med.	High	High	Med.	Table Dessert/Dessert and Med. sweet

Table 5 cont	Tannin	Acid	Sugar	Pectin	Flavour	Body	Wine style suitability
Red-currants	Med.	High	Med.	High	Med.	Med.	Table/Med. sweet
Strawberries	Low	Med.	High	Low	Med.	Low	Table dessert/Dessert and Med. sweet
Tangerines	Med.	High	Med.	Med.	Med.	Med.	All wine styles
Dried: Bilberries	High	Med.	Med.	Med.	High	High	Table/Med. sweet/Dessert
Elderberries	High	Low	Low	Med.	High	Med.	Table/Med. sweet/Dessert
2. Fruits which are best mixed with other ingredients							
Fresh: Bananas	Low	Low	Med.	Low	Med.	High	All wine styles
Crab apples	High	High	High	Low	Med.	Low	Aperitif/Table/Medium sweet
Lemons	Low	High	Low	High	Med.	Med.	Aperitif
Pears	High	Low	Med.	Low	High	Low	Table/Med. sweet
Pineapple	Low	High	Med.	High	Low	Low	Table/Med. sweet
Rhubarb	Low	High	Med.	Med.	Med.	Low	Table/Med. sweet
Sloes	High	High	Med.	High	High	Med.	Table/Med. sweet/Dessert
Dried: Bananas	Low	Low	High	Low	High	High	All wine styles
Dates	Low	Low	High	Low	High	High	Dessert
Currants	Med.	High	High	Med.	Med.	Med.	All wine styles
Figs	Low	Low	High	Med.	High	High	Dessert
Raisins	High	Med.	High	Med.	High	High	Dessert
Sultanas	Med.	Med.	High	Med.	Med.	High	All wine styles
Prunes	Med.	Med.	High	Med.	High	High	Dessert
Rose hip shells/ Rose hips	High	Low	High	Med.	Med.	High	All wine styles
3. Other ingredients which are very useful additives							
Grape concentrate	Med.	Med.	High	Med.	Med.	High	All wine styles
Honey	Low	Low	High	Low	Med.	High	All wine styles
Flowers	Low	Low	Low	Low	Low	Low	All wine styles—for bouquet

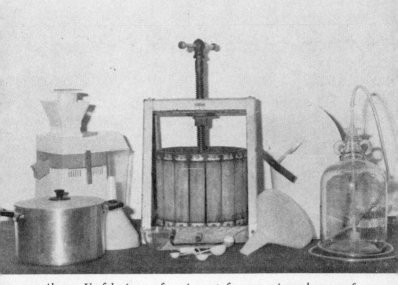

Above: Useful pieces of equipment for any winemaker ... for juice extraction, an electric juice extractor (back left) a small fruit press, and a Pulpmaster attachment and bucket for use with an electric drill. In front, a stainless steel saucepan, funnels, Oetker balance and measuring spoons, and a fermentation jar and siphon with tap.

Below: Chemicals widely used by the winemaker ... citric acid, ammonium phosphate, hydrogen phosphate, ammonium sulphate, wine tannin, Vitamin B₁ (Benerva), yeast nutrient, and various liquid and tablet wine yeasts.

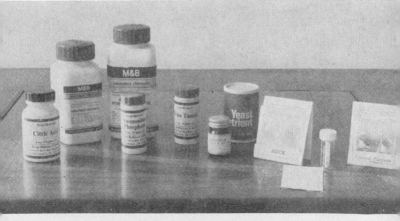

Titrating a wine to check its acidity, and whether it needs adjustment (see Page 235).

Pectin testing reagent is used to discover whether pectin is the cause of a wine haze.

Above: Testing the specific gravity of a finished wine by means of the hydrometer ...

Below: ... and measuring the amount of residual sugar by means of a Clinitest outfit.

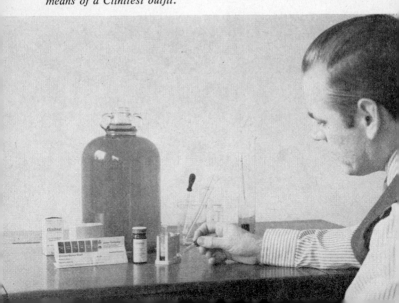

Using the foregoing tables as a guide, it should now be possible to formulate individual recipes. For example, if one wishes to make a red table wine to accompany the main course of a meal, reference to the tables will show the ingredients which are suitable and the acidity, alcohol content, depth of flavour and body at which to aim.

As a starting point for the formulation of recipes, general details relating to the ingredients, their preparation and "tried and tested" recipes are now given.

These are for one gallon lots in every case but quantities can naturally be increased for multiples of a gallon, if required.

Ammonium sulphate is mentioned in all instances as the yeast growth promoter but ammonium phosphate and thiamine may also be added if the reader wishes after evaluating the remarks concerning these in chapter 3.

Where astringency is required the use of grape tannin is recommended.

The yeast starter bottle should be prepared two or three days before the fruit is processed so that no delay occurs when the yeast is needed. The choice of yeast to be used is a matter of personal preference. As explained earlier in the book, wine yeast cultures on agar slants or in liquid form are preferable to other forms of yeast. The style of yeast, i.e. Sauternes, Tokay, Champagne, general purpose, etc., is also a matter for individual choice and experiment. Different strains will produce slightly different results and flavours which are particularly noticeable with light-flavoured wines and some yeast strains are more alcohol-tolerant than others; so choose a sympathetic yeast, where possible, for the style of wine to be made but this is by no means essential. The differences in flavour under home winemaking conditions are often so slight that a good strain of wine yeast will prove quite satisfactory for varying styles of wine.

Extension of the fruit with water will be necessary in many instances and a guide to volume extensions are given in chapter 3. However, with the inclusion of several ingredients in a "must" and "pulp" fermentation, it is often easier to use about five pints of water with the fruit and to top up to a

127

gallon after the sugar and acidity corrections have been made. This method will be clear from the recipes.

"Pulp" fermentation is recommended in most instances in order to extract the full flavour and sweetness, tannin, pigments, etc., from the fruit and this should be carried out in a container suited to the quantity of wine being made, full enough to allow vigorous ferment to proceed but, at the same time, closely covered, leaving as little air space as possible to avoid contamination in the initial phase between preparation and fermentation.

One of the quickest ways to strain the juice from the "pulp" into the fermentation jar is to use a funnel and polythene strainer. Have close at hand another sterilised wide-necked jar with a gallon polythene funnel inserted and, inside the funnel, a sterilised jelly bag or other suitable straining bag. Take the "pulp" and let the juice run freely through the strainer into the fermentation jar, and as soon as the "pulp" begins to clog, move to the other wide-necked jar and place the "pulp" in the straining bag. (It will be necessary to rinse the strainer at intervals). At the end of this process, the bulk of the juice will be in the fermentation jar, to which may be fitted an air-lock and the "pulp" will be left dripping through the straining bag, which will take several hours. One gram of sodium metabisulphite should be added to this juice as an anti-oxidant, in view of the air space in the jar and the length of time involved. Also, the gallon funnel containing the straining bag should be securely covered to exclude unwanted air. A large sheet of thick plastic with thin elastic to secure, will suffice. When all the juice appears to have run through, this may be added to the fermentation jar.

This method is easy and obviates a lot of messy hand-squeezing of the "pulp."

Racking of the wine two or three times in the first year after fermentation, fining if necessary and the length of time for maturation is left to the discretion of the reader.

128

Some of the weights given in this chapter are quoted in grammes because the amount needed is too small to be given in ounces. The gramme is a weight commonly used in scientific work and all over the Continent and many readers will be familiar with it, but for those who are not it is approximately one twenty-eighth of an ounce. It will be found that a 1½ in. oval nail is an excellent weight for this amount!

To enable weighing of ½, 1, 2 and 3 gm. quantities only two weights need be purchased, viz. ½ and 1 gm. 2 gms. can be weighed out by taking two 1 gm. quantities and 3 gms. using two 1½ gm. amounts.

Making a simple balance

If a simple balance is not available one can easily be made from 2 tin lids, 2 knitting needles, some thin twine, a medicine bottle and some sticky paper.

Technique of making balance

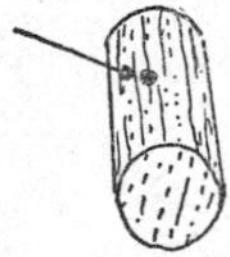

Hole

Take a round cork of approximately (12.5 mm.) diameter and (2.5 cms.) long and a plastic knitting needle of 10 gauge.

Drill a hole slightly smaller than the needle, in the middle of the cork.

Push the needle through the hole so that the cork is about central. Cut off the head of the needle with a hack saw, file the needle into a slight point and check the cork is central (fig. 2).

Fig. 1

Fig. 2

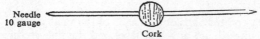

Needle 10 gauge

Cork

Next take two small corks (9.5 mm.) diameter and (15.9 mm.) long approximately and fit one to each end of the needle

Fig. 3

Now take a very fine knitting needle and trim it with a hack saw or tin snips so that it is (2.5 cms.) shorter than the

medicine bottle height. Push this needle firmly into the middle cork so that it is central but at right angles to the 10 gauge needle.

Insert the fine needle into the bottle so that the large cork sits on the neck (fig. 4).

Fig. 4

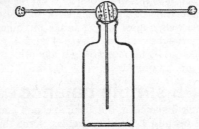

Using two tin lids of the same size (7.5 cms.) in diameter is ideal), mark a cross on the bottom of each (fig. 5) with a pencil and punch a small hole in the rim above each of the cross ends (fig. 6).

Fig. 5

Fig. 6

Holes for twine

Take four (23 cms.) lengths of twine and knot one end of each. Use a piece to join holes 1 and 2 (see fig. 5) and another to join holes 3 and 4. Tie the ends so that both loops on each lid are the same size.

Now suspend the pans one on each of the small corks.

The balance should now rest level and the fine knitting needle should be perpendicular. If this is not so and the balance goes down to the left, slide the left hand cork and pan in slightly towards the middle so that balance is obtained. If this is ineffective a little melted candle wax can be added to whichever pan is light.

All that then remains is to stick a strip of adhesive paper on the outside of the bottle and mark the balance point of the needle on it. The finished product is shown in fig. 7.

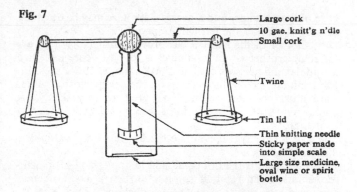

Fig. 7

- Large cork
- 10 gae. knitt'g n'dle
- Small cork
- Twine
- Tin lid
- Thin knitting needle
- Sticky paper made into simple scale
- Large size medicine, oval wine or spirit bottle

RECIPES

The following recipes have the correct percentage acidities which are meant to be achieved by titration and adjustment (p. 29 and p. 235). However, to assist those without test equipment, I have shown at the end of each list of ingredients the approximate weight in grammes of citric acid which might be used. It must be stressed that because of variation in the acidity of the various fruits this is only an approximation and that the titration method is to be preferred when ever possible.

Apple

There are many varieties and a note of the types used, where known, will be most useful for guidance when preparing future recipes. A good plan is to use several varieties, both eating and cooking, including some crab apples if available, to add some astringency. Smell the various types and make a note of those which have the best aroma.

A light bodied wine is invariably produced and, if the acidity is right, the juice only may be fermented with no extension, only sugar adjustment. Of course, if more body is required in the wine, additional ingredients may be included to provide this.

Crushing or electric juicer extraction is ideal but care must be taken to sulphite immediately, as apples oxidise very easily. After standing for a few hours, solids will fall to the

131

base of the container and a nearly clear juice may be decanted or racked off into the fermentation vessel.

Another method which has proved satisfactory, where a juicer or crusher is not available, is to wash, core and slice the apples and pressure cook in a little water at 15 lbs. pressure for 1 minute, before fermenting on the "pulp" for 2/3 days. Care must be taken to bring to pressure very slowly, as apples tend to froth up during cooking and may block the vent tube.

It is interesting to evaluate the flavours obtained by the different methods.

Malic acid is the principal acid in apples and malolactic fermentation can result if sulphur dioxide is not prudently employed to prevent bacterial infections. This is fully explained in chapter 12.

Apple: *Wine style, Table dry*

Aim:

Flavour	Light
Body	Light
Bouquet	Light/medium
Approx.% alcohol	12
Starting S.G.	1.088
Final S.G.	Approx. .990
Acidity	0.7% as citric

Ingredients:

Working yeast starter	1 gm. tannin, if no crab
5 pints (3 litres) juice from	apples used
Pectic enzyme	3 gms. ammonium sulphate
1½ gms. sodium metabisulphite	Sugar and acid to adjust

If no estimation of acidity is made use 20 gms of Citric acid.

Method:

Prepare the juice by electric extractor or crushing and put in a sterile container with 1 gm. sodium metabisulphite and pectic enzyme. Stopper. After a few hours decant supernatant juice into clean container and take specific gravity reading. Adjust sweetness to 1.088. Take acidity reading and adjust to 0.7% as described in chapter 3. Add tannin, ammonium sulphate and working yeast starter as described in chapter 4. When yeast is working well in gallon jar add boiled, cool water, if necessary, to top up to one gallon. Fit air-lock and leave to ferment to dryness. When fermentation ceases, place jar in refrigerator for three days. Rack upon removal. Add ½ gm. sodium metabisulphite. Rack and fine as necessary. Store to mature.

Apple: *Wine style, Medium sweet*
(*To drink with savouries*)

Aim:

Flavour	Medium
Body	Medium
Bouquet	Medium
% alcohol	14
Starting S.G.	1.100
Final S.G.	.998/1.000—arrested fermentation
Acidity	0.65% as citric
Alternatively:	
Starting S.G.	1.090
Final S.G.	.990
	Sweeten to 0.998/1.000

Ingredients:

Working yeast starter

Pectic enzyme

6–8 lb. (2¾–3½ kilo) apples, mixed

1 gm. tannin, if no crab apples used

1 lb. (½ kilo) sultanas

(alternatively 1 pint grape white concentrate)

1½ gms. sodium metabisulphite

3 gms. ammonium sulphate

1 lb. (½ kilo) fresh bananas

1 gm. potassium sorbate
Sugar and acid to adjust

If no estimation of acidity is made use 17 gms Citric acid.

Method:

Pressure cook prepared apples and bananas. Empty into sterile container and cover. If sultanas are added, clean, pressure cook and add to apples and bananas. If concentrate is used, this may be melted in the hot liquid after pressure cooking the other ingredients. Add cool boiled water to bring bulk almost up to one gallon. Cover. When cool, add 1 gm. sodium metabisulphite, pectic enzyme, ammonium sulphate and tannin. Stir well, cover and leave for six hours approximately. Strain off enough juice to take specific gravity reading and adjust sweetness to 1.098. Take acidity reading and adjust to 0.65%. Add yeast and ferment on the "pulp" for two/three days, keeping well covered and stirring night and morning. Strain into fermentation jar and top up with cool, boiled water if necessary. Fit air-lock and let fermentation proceed to .998/1.000. (Test with hydrometer when fermentation slows right down). Place jar in refrigerator for three days. Rack. Add 1 gm. potassium sorbate to stabilise and ½ gm. sodium metabisulphite. Rack and fine, as necessary. Store to mature.

Apricot

These are fairly costly to buy fresh but tinned and dried apricots are available all the year round, although the latter tend to produce a high concentration of gums in the "must." The flavour of apricots blends well with nearly all other fruits.

Fresh and tinned apricots may be pulped or liquidised, as mentioned in chapter 3. Alternatively, if the fresh apricots are rather hard, boiling water may be poured over to soften before pulping. As many stones as possible should be removed, although these will be left at the bottom of the fermenting vessel after two/three days of "pulp" fermentation.

Dried apricots should be pressure cooked for five/six minutes at 15 lb. pressure in a little water, about ½ pint per 1 lb. of fruit, or using the boiling water technique used as explained in chapter 3.

Apricot: *Wine style—Dessert: Fortified*

Aim:

Flavour	Full
Body	Full
Bouquet	Full
% alcohol	18–20
Starting S.G.	1.100
Final S.G.	Approx. .990
	Sweeten to 1.038
Acidity	0.50% as citric

Ingredients:

Working yeast starter
4 lb. (1¾ kilos) fresh apricots
2 lb. (1 kilo) fresh bananas
Pectic enzyme
1 gm. tannin

1 gm. sodium metabisulphite
1½ gms. ammonium sulphate
Sugar and acid to adjust
Vodka or other neutral-
flavoured spirit

If no estimation of acidity is made use 23 gms Citric acid.

Method:

Liquidise the apricots and bananas and extend volume of fruit one and a half times with boiled water. When cool, add 1 gm. sodium metabisulphite, ammonium sulphate and pectic enzyme. Also tannin. After about six hours strain off juice sample and test for specific gravity, adjust to 1.100. Determine acidity and adjust to 0.50%. Add yeast starter and, if necessary, top up gallon fermenting vessel with cool boiled water so that only a little air space is left. Stir well, cover and ferment on the "pulp" for two days, stirring night and morning and keeping well covered between stirrings. Strain into fermenting vessel. Top up jar with boiled cool water if necessary and fit air-lock. Ferment out completely. Place in refrigerator for three days. Rack. Determine alcohol content as mentioned in chapter 16. Sweeten with sugar to 1.038 specific gravity and fortify to 18–20% alcohol, with neutral spirit as described in chapter 6. Refrigerate for one week. Rack into clean container. Rack and fine as necessary. Store to mature.

Apricot: *Wine style, Table dry*

Aim:

Flavour	Light
Body	Light
Bouquet	Light/medium
% alcohol	12 approx.
Starting S.G.	1.088
Final S.G.	Approx. .990
Acidity	0.7% as citric

Ingredients:

Working yeast starter
½ kilo tinned apricots
½ pint (230 mls.) white grape concentrate
½ lb (¼ kilo) sultanas
Pectic enzyme

½ gm. tannin
1½ gm. sodium metabisulphite
1½ gm. sodium metabisulphite
Sugar and acid to adjust

If no estimation of acidity is made use 32 gms Citric acid.

Method:

Liquidise apricots and extend with one and half volumes boiled cooled water. Alternatively extend with boiled cooled water and macerate with clean hands. Melt concentrate in boiling water. Clean and mince sultanas and add to other ingredients or pressure cook in ½ pint water and add to other ingredients. When cool add pectic enzyme, metabisulphite, (1 gm.), ammonium sulphate and tannin. After steeping for approximately six hours strain off sample and test specific gravity and adjust to 1.088. Test acidity and adjust to 0.7%. If necessary, make bulk up to almost a gallon with cool boiled water, add yeast starter and ferment for two/three days on "pulp" in covered container, stirring night and morning. Strain into fermentation jar as mentioned previously, topping up with boiled cool water if necessary and fit air-lock. Ferment to dryness. Refrigerate for three days. Rack and add ½ gm. sodium metabisulphite. Rack and fine, as necessary. Store to mature.

Banana

This fruit is known to add body to a wine and is a very useful additive to many fermentations for this reason and also for the fact that another attribute of the banana is that brilliantly clear wines invariably result when they are used. To give body, approximately 2 lbs. per gallon may be used for a dessert style with lesser amounts for lighter bodied wines. A useful tip is to use the liquid from a few simmered or pressure cooked sliced and peeled bananas to top up jars after racking, if too much air space is left.

They may be peeled and liquidised for inclusion in a "pulp" fermentation of short duration but the best method of preparation is to pressure cook at 15 lbs. pressure for five minutes in a little water, or simmer for 20/30 minutes in an open pan in a little water. The liquid may then be strained off and added to the "must."

The dried variety can be used if desired but these sometimes impart an odd flavour to the wine and with the availability of fresh fruit at a reasonable price all the year round, there seems no advantage in using these.

Banana, prune and date: *Wine style, Dessert: Fortified*

Aim:

Flavour	Full
Body	Full
Bouquet	Full
% alcohol	18–20
Starting S.G.	1.100
Final S.G.	Approx. .990
	Sweeten to 1.038
Acidity	0.50% as citric

Ingredients:

Working yeast starter
  (Sherry yeast if possible)
½ lb. (¼ kilo) prunes
3 lb. (1¼ kilos) bananas
½ lb. (¼ kilo) dates, stoned
1 gm. tannin

Pectic enzyme
3 gms. ammonium sulphate
1½ gms. sodium metabisulphite
Sugar and acid to adjust
Vodka or other neutral-
  flavoured spirit

If no estimation of acidity is made use 23 gms Citric acid

Method:

Peel and slice the bananas and pressure cook for five minutes at 15 lbs. pressure in a little water, or simmer for 20/30 minutes in three pints water. Pressure cook prunes and chopped dates for 10–15 minutes in 1 pint of water at 15 lbs. pressure or simmer with bananas in 4 pints of water in all. Place in sterile container, cover and when cool add tannin, pectic enzyme, sodium metabisulphite and ammonium sulphate plus enough cool boiled water to make the bulk to almost a gallon. Stir well and leave for six hours approximately. Strain off sample of the liquid and test for specific gravity. Adjust with sugar to 1.100. Test for acidity and adjust to 0.50%. Add working yeast starter and leave to ferment for approximately three days on the "pulp," keeping well covered in between stirring night and morning. Strain into sterile fermentation jar, top up with cool boiled water if necessary, fit air-lock and leave to ferment to dryness. When fermentation is complete, place in refrigerator for three days. Rack. Determine alcohol content as mentioned in chapter 16. Sweeten with sugar to 1.038 and fortify to 18–20% alcohol as described in chapter 6, with neutral spirit. Refrigerate for one week. Rack and fine, as necessary. Store to mature.

Bilberry

These are a good medium particularly for table wines and they are mainly available in dried and tinned forms. Also the popularity of growing this fruit in gardens is increasing with winemakers.

Dried bilberries require careful selection at the time of purchase. Smell them and if there is an unpleasant musty aroma, purchase elsewhere or wait until a new batch is available. As with all dried fruit, sterilisation is most important, the micro-organism population being high. Dried bilberries should be thoroughly rinsed and, if possible, pressure cooked for 10–15 minutes at 15 lbs. pressure in a little water. Alternatively the boiling water and sulphite technique may be used as explained in chapter 3. Pulp fermentation is desirable to extract the full flavour from the fruit.

Bilberry: *Wine style, Table dry*

Aim:

Flavour	Full
Body	Medium
Bouquet	Full
% alcohol	12
Starting S.G.	1.088
Final S.G.	Approx. .990
Acidity	0.57% as citric

Ingredients:

Working yeast starter	3 gms. ammonium sulphate
¾ lb. (350 gms.) bilberries, dried	Acid to adjust
1 lb. (½ kilo) bananas	Sugar to adjust
Pectic enzyme	1½ gms. sodium metabisulphite

If no estimation of acidity is made use 26 gms Citric acid.

Method:

Rinse the bilberries very thoroughly and pressure cook at 15 lbs. pressure for 15 minutes in approximately 1 pint of water. Place in a sterile container. Peel and slice the bananas and pressure cook in a little water for five minutes at 15 lbs. pressure. Add to bilberries in container. Add cool, boiled water to make up bulk to approximately ¾ gallon. When cool add 1½ gms. sodium metabisulphite, pectic enzyme, 3 gms. ammonium sulphate and leave, well covered for about six hours. Take sample of liquid and determine specific gravity. Adjust with sugar to 1.088. Test acidity and adjust to 0.57%. Add working yeast starter and ferment on the "pulp" for four days. Strain, as mentioned previously, into fermentation jar, top up with cool boiled water if necessary and fit air-lock. Ferment to dryness. Place in refrigerator for three days. Rack. Rack again and fine, as necessary. Store to mature.

Blackberry

The joy of this commodity is that it can be obtained from the hedgerows for nothing and an outing picking blackberries adds to the pleasure of winemaking for many.

Blackberries should be washed and carefully inspected for "unwanted inhabitants" as it is not desired to add body to blackberry wine with maggots! Sterilisation of the "must" again cannot be over-emphasized as failure to attend to this detail produces "off" flavoured wines.

Preparation is by pulping or liquidising.

Blackberries and apples are a natural combination and a recipe is given below.

Blackberry and apple: *Wine style, Medium sweet*
(*To drink with savouries*)

Wine style	*Medium Sweet*
(*To drink with savouries*)	

Aim:

Flavour	Medium
Body	Medium
Bouquet	Medium
% alcohol	12–14
Starting S.G.	1.098
Final S.G.	.990
	Sweeten to .998/1.000
Acidity	0.65% as citric

Ingredients·

Working yeast starter	Pectic enzyme
3 lb. (1½ kilos) blackberries	1 gm. sodium metabisulphite
3 pints (1¾ litres) apple juice	3 gms. ammonium sulphate
½ lb. (¼ kilo) sultanas	1 gm. potassium sorbate
1 lb. (½ kilo) bananas	Sugar and acid to adjust

If no estimation of acidity is made use 23 gms Citric acid.

Method:

Wash blackberries and examine carefully. When quite clean, place in a sterile container. Liquidise blackberries if possible. Add 1 gm. sodium metabisulphite. Add 3 pints apple juice (see under "Apples" for details). Wash and mince sultanas, peel and slice bananas and pressure cook or simmer these ingredients in a little water (five minutes at 15 lbs. pressure or 20–30 minutes in an open pan). Add to other ingredients in container. Add pectic enzyme, ammonium sulphate and enough cool boiled water to bring bulk up to approximately ¾ gallon. Stir well and cover for six hours approximately. Strain off sample of juice and test for specific gravity and adjust with sugar addition to 1.098. Test acidity and adjust to 0.65%. Add yeast and ferment on the "pulp" for two/three days. Strain into fermentation jar, top up with cool boiled water and fit air-lock. Ferment to dryness. Place in refrigerator for three days. Rack. Sweeten to .998/1.000 and add 1 gm. potassium sorbate. Rack and fine, as necessary. Store to mature.

Blackcurrant

Unless one has a large garden with several bushes of blackcurrants, these can be an expensive commodity to buy fresh, but fortunately, due to the strong flavour, only a small quantity is needed. Remove the stalks but there is no need to "top" the fruit.

The skins are tough so if one does not possess a liquidiser, boiling water poured over the fruit will help the softening operation. Another method with fruit of this nature is to put it into the freezer section of the refrigerator for two or three days, when the fruit thaws, the juice will run freely. Pressure cooking at 15 lbs. pressure for three minutes is another method but this can give a tawny wine instead of a red one, a recipe for which is given. When pressure cooking fruit, remember the flavour is slightly altered and only a little water is required. However, it has the advantage of completely sterilising the fruit, whilst softening ready for fermentation, in a matter of minutes. Additional cool water can be added immediately after cooking to make up the necessary volume, thus obviating a slow cooling down period, as with the boiling water method.

The reader may wish to experiment by making two wines from the following recipe, one with liquidising and extending with cool boiled water and the other by pressure cooking the fruit. Take particular note of the flavour and colour variations.

Blackcurrant: *Wine style, Dessert: Unfortified*

Aim:

Flavour	Full
Body	Full
Bouquet	Full
% alcohol	15
Starting S.G.	1.100
Final S.G.	Approx. .990
	Sweeten to taste
Acidity	0.50% as citric

Alternatively, if desired, this wine
may be fortified as mentioned
under other dessert types.

Ingredients:

Working years starter
1½ lb. (700 gms.) black-
currants
¾ lb. (350 gms.) stoned raisins
2 lb. (1 kilo) bananas
Pectic enzyme

Sugar and acid to adjust
1½ gm. sodium metabisulphite
3 gms. ammonium sulphate
1 gm. potassium sorbate

If no estimation of acidity is made use 2 gms Citric acid.

Method:

Prepare blackcurrants by method of choice, also bananas
and raisins and put all together in sterile container with 1½
gms. sodium metabisulphite, pectic enzyme and 3 gms. ammonium sulphate. Stir well and cover. After about six
hours strain off sample, test for specific gravity and adjust to
1.100. Test acidity and adjust to 0.50%. Add working yeast
and top with cool boiled water to almost a gallon. Stir well
night and morning and keep well covered in between stirring.
After fermentation has proceeded for about three days,
strain into fermentation jar. Top up jar with boiled cool
water and fit air-lock. Ferment to dryness. Place in refrigerator for three days. Rack. Sweeten to taste as 1.038 S.G.
may be a little sweet for an unfortified wine. Add 1 gm.
potassium sorbate. Store. Rack and fine, as necessary.
Store to mature.

Cherry

Cherry wine has a delightful flavour if made in accordance with the principles of this book and sometimes has an "almondy" characteristic from fermentation for a few days on the fruit and stones. In fact, a few stones may be cracked and the kernels put into the fermentation if this flavour is wanted. Boiling water is again useful for softening the skins or the fruit may be steeped in water with sulphite and the pectic enzyme, and macerated after several hours. A mixture of sweet and cooking cherries is desirable but not essential.

Cherry: *Wine style, Dessert: Unfortified*

Aim:

Flavour	Full
Body	Full
Bouquet	Full
% alcohol	15
Starting S.G.	1.100
Final S.G.	Approx. .990
	Sweeten to taste
Acidity	0.45% as citric

Alternatively, if desired, this wine may be fortified as mentioned under other dessert types to 18–20% alcohol.

Ingredients:

Working yeast starter
3 lb. (1½ kilo) or more cherries, black or Morello or mixed
1 pint (½ litre) red grape concentrate
2 lb. (1 kilo) bananas

Pectic enzyme
1½ gm. sodium metabisulphite
3 gms. ammonium sulphate
1 gm. potassium sorbate

Sugar to adjust

Method:

Wash the cherries and steep in water, using whichever method is preferred and when cool add 1 gm. sodium metabisulphite, pectic enzyme, and 3 gms. ammonium sulphate. Melt concentrate in water and leave in separate container with $\frac{1}{2}$ gm. sodium metabisulphite for approximately six hours. Take separate samples of cherries and concentrate and titrate for acidity. As only a low acidity is required, consideration will then have to be given as to the quantity of cherries which may be added to make up the correct level of acidity Add as many as the acidity will allow and macerate. Add the concentrate. Peel and slice the bananas and add these or the cooled liquid from them to the other ingredients. Stir well and take another sample from the mixed ingredients. Test for specific gravity and adjust to 1.100. Test also for acidity, to see if the original calculations are correct. If necessary, adjust to 0.45%. Add yeast starter and ferment for two/three days. Strain into sterile fermentation jar, top up with cool boiled water, fit air-lock and leave to ferment to dryness. When fermentation is complete, place in the refrigerator for three days. Rack. Sweeten to taste and add 1 gm. potassium sorbate for the unfortified wine. Store, rack and fine, as necessary. Alternatively after the first racking, determine alcohol content, sweeten to 1.038 and fortify, as explained in chapter 6, to 18–20% with a neutral-flavoured spirit. Refrigerate for one week. Rack again and store.

Cherry: *Wine style, Table dessert*

Aim:

Flavour	Medium
Body	Medium
Bouquet	Full
% alcohol	14
Starting S.G.	1.100
Final S.G.	Approx. .990
	Sweeten to 1.020
Acidity	0.7% as citric

Ingredients:

Working yeast starter	1½ gms. sodium metabisulphite
4 lb. (1¾ kilo) cherries	Pectic enzyme
½ pint (¼ litre) red grape concentrate	3 gms. ammonium sulphate
1 lb. (½ kilo) bananas	1 gam. potassium sorbate

If no estimation of acidity is made use 30 gms Citric acid.

Method:

Prepare ingredients and extract flavour, sweetness and acidity as detailed above. After six hours test for specific gravity and acidity and adjust to 1.100 and 0.7% respectively. Top up container to almost a gallon with cool boiled water and add yeast starter. Ferment for two/three days, stirring night and morning, otherwise keeping well covered. Strain in the usual way into fermenting jar and top up with cool boiled water. Fit air-lock. Ferment to dryness. Refrigerate for three days. Rack. Add sugar to sweeten to taste or to 1.020 and also add 1 gm. potassium sorbate to stabilise. Rack and fine, as necessary. Store to mature.

Cherry: *Wine style, Table dry*

Aim:

Flavour	Full
Body	Medium
Bouquet	Full
% alcohol	12
Starting S.G.	1.088
Final S.G.	Approx. .990
Acidity	0.57% as citric

Ingredients:

Working yeast starter	1 gm. sodium metabisulphite
2–3 lb. (1¼ kilo) cherries	Pectic enzyme
½ lb. (¼ kilo) bananas	1 gm. tannin
½ lb. (¼ kilo) sultanas	3 gms. ammonium sulphate

If no estimation of acidity is made use 26 gms Citric acid.

Method:

Process in the same way as for "table dessert" but adjust specific gravity to 1.088 and acidity to 0.57%. Ferment on the "pulp" for two/three days. Strain into fermentation jar and ferment to dryness. Refrigerate for three days. Rack and fine, as necessary. Store to mature.

Damson

This fruit is a fine medium for dessert wines and is also excellent combined with sloes which ripen at about the same time. Here again, steeping is necessary to soften the fruit before pulping. The acidity of the fruit often determines the quantity to be used.

Damson: *Wine style, Dessert*

Aim:

Flavour	Full
Body	Full
Bouquet	Full
% alcohol	15 unfortified
	18–20 fortified
Starting S.G.	Approx. 1.100
Final S.G.	Approx. .990
	Sweeten to taste or to 1.038
Acidity	0.45% as citric

Ingredients:

Working yeast starter
3 lb. (1½ kilo) damsons (or more)
or
2 lb. (1 kilo) damsons
1 lb. (½ kilo) sloes

1 lb. (½ kilo) sultanas or 1 pint red grape concentrate
2 lb. (1 kilo) bananas

Pectic enzyme
1½ gms. sodium metabisulphite

3 gms. amonium sulphate
Sugar to adjust

1 gm. potassium sorbate if wine is left unfortified

If no estimation of acidity is made use 6 gms Citric acid.

Method:

Prepare ingredients and proceed as for Cherry Dessert and make specific gravity and acidity corrections to 1.100 and 0.45% respectively. Ferment on "pulp" for approximately four days. Strain into fermentation jar and top up with cool boiled water. Fit air-lock. Ferment to dryness. Refrigerate for three days and rack. Sweeten to taste if wine is to be left unfortified and add 1 gm. potassium sorbate. Alternatively after racking, determine alcohol content, sweeten to 1.038 and fortify to 18–20% alcohol. Refrigerate for one week. Rack and fine, as necessary. Store to mature.

Damson: *Wine style—Table dry*
Aim:

Flavour	Full
Body	Medium
Bouquet	Full
% alcohol	12
Starting S.G.	1.088
Final S.G.	Approx. .990
Acidity	0.57% as citric

Ingredients:

Working yeast starter
2 lb. (1 kilo) damsons
1 pint (½ litre) red grape concentrate or 1 lb. (½ kilo) sultanas

Pectic enzyme
1½ gm. sodium metabisulphite
3 gms. ammonium sulphate
Sugar and acid to adjust

If no estimation of acidity is made use 9 gms Citric acid.

Method:

Prepare ingredients and proceed with specific gravity and acidity adjustments in usual way. Add working yeast starter and ferment on "pulp" for about four days. Strain into fermentation jar and fit air-lock. Ferment to dryness. Refrigerate three days and rack. Rack again and fine, as necessary. Store to mature.

Elderberry

There are many types of elderberries, both red and white, round and oval with green or red stems and it is said the varieties most favoured by birds (of the flying variety) are the ones most suitable for winemaking—but one has to beat the birds! Make a point of marking the best spots for elderberry bushes on a road map and note the origin in the recipe book for future reference.

As mentioned previously, elderberries are excellent for a dessert or table style but it must be remembered that the flavour is strong and more berries will be required for a dessert wine than for the table style. (As a general guide, 4 lbs. and 2 lbs. respectively).

The berries may be removed from the stalks with a fork, ready for liquidisation, or steeping in water (boiling water again here will soften the skins).

Alternatively the berries may be pressure cooked at 15 lbs. for three minutes. A useful method here is to use more berries than usual and strain off the "pulp" immediately, ready for fermentation. This will give the colour and flavour required by using a higher quantity of fruit but no bitterness is extracted from the skins.

With "pulp" fermentation, flavour, tannin and colour will be extracted due to the alcohol so produced but an unpleasant bitterness often results if left too long on the skins.

Although the tannin content is high, acidity is low and must be carefully adjusted to obtain a clean-flavoured wine.

Dried elderberries are also obtainable but these are not nearly so good as fresh elderberries. If dried elderberries are used, the same remarks apply as for dried bilberries. A recipe is given below using dried elderberries and bilberries, which has proved very successful, but which follows a slightly different procedure to those already mentioned in that it utilises the same ingredients for a gallon of dessert wine and a gallon of table wine with a "second mashing" technique.

Elderberry and bilberry: dried fruit: *Wine styles—Dessert and table*

Aim:

Flavour	Full	Full
Body	Full	Medium
Bouquet	Full	Full
% alcohol	15 or 18–20	12
Starting S.G.	1.100	1.088
Final S.G.	.990	.990
	Sweeten to taste	
Acidity	0.45%	0.57% as citric

Ingredients:

Working yeast starter	3 gms. sodium metabisulphite
½ lb. (¼ kilo) dried elderberries	6 gms. ammonium sulphate
½ lb. (¼ kilo) dried bilberries	Sugar and acidity to adjust
½ lb. (¼ kilo) sultanas	1 gm. potassium sorbate if dessert gallon is left unfortified
2 ozs. (60 gms.) figs	
1 lb. (½ kilo) honey	
2 lb. (1 kilo) bananas	Neutral spirit if dessert gallon is fortified
Pectic enzyme	

If no estimation of acidity is made for 0.45% use 20 gms

148

Citric acid or for 0.57% 25 gms Citric acid.

Method:

In the first instance, proceed to make the dessert wine. Prepare the ingredients by the method preferred, add 1½ gms. sodium metabisulphite, 3 gms. ammonium sulphate and pectic enzyme. Adjust the specific gravity and the acidity. Place in fermentation vessel and ferment on "pulp" for two/three days. No longer. Strain off juice but do not place "pulp" in straining bag to extract further juice. Instead, put "pulp" back immediately in original container, top with cool boiled water to almost a gallon, add 1½ gms. sodium metabisulphite, 3 gms. ammonium sulphate, cover and leave while completing the dessert gallon preparation.

Top up juice in fermentation jar of dessert gallon with cool boiled water if necessary. Fit air-lock and let fermentation proceed to dryness. If an unfortified dessert is required, refrigerate, rack, sweeten to taste, add 1 gm. potassium sorbate. Rack again and fine if necessary. Store to mature. If a fortified dessert is required, after the initial racking, determine alcohol content from details in chapter 16 "Methods of Test," sweeten to 1.038 and fortify with a neutral-flavoured spirit. Refrigerate for one week and rack again. Fine and rack again as necessary. Store to mature.

As soon as the air-lock is fitted in the dessert gallon return to the table gallon in the original fermentation vessel, stir well and leave for about six hours. Take specific gravity and adjust to 1.088, test acidity and adjust to 0.57%. Ferment on "pulp" for a further three/four days. Strain and proceed in the usual way for a table wine, fermenting to dryness.

Elderberry: fresh: *Wine style, Dessert*

Aim:

Flavour	Full
Body	Full
Bouquet	Full
% alcohol	15 unfortified
	18–20 fortified
Starting S.G.	1.100
Final S.G.	.990
	Sweeten to taste or 1.038
Acidity	0.45% as citric

Ingredients:

Working yeast starter	Pectic enzyme
4 lb. elderberries	1½ gms. sodium metabisulphite
2 lb. bananas	3 gms. ammonium sulphate
1 pint grape concentrate	Sugar and acid to adjust
(or 1 lb. raisins, stoned)	

If no estimation of acidity is made use 15 gms Citric acid.

Method:

Proceed generally in accordance with other deserts wines. Ferment on "pulp" for three days.

Elderberry, fresh: *Wine style—Table*

Aim:

Flavour	Full
Body	Medium
Bouquet	Full
% alcohol	12
Starting S.G.	1.088
Final S.G.	.990
Acidity	0.57% as citric

Ingredients:

Working yeast starter	1½ gm. sodium metabisulphite
2 lb. (1 kilo) elderberries	3 gms. ammonium sulphate
1 lb. (½ kilo) sultanas	Sugar and acid to adjust
1 lb. (½ kilo) bananas	

If no estimation of acidity is made use 23 gms of Citric acid.

Method:

Proceed as for other table wines and ferment on "pulp" for three days.

Gooseberry

These should be picked or purchased at the peak of condition and a mixture of both cooking and dessert types is preferable. A gooseberry aroma is a deterrent to some people but can be pleasant in a light table wine.

Preparation is by liquidisation, steeping in water or pressure cooking up to 15 lbs. pressure only. There is no need to top and tail the fruit.

Gooseberry: *Wine style, Table (Sea food and fish)*

Aim:

Flavour	Light
Body	Light
Bouquet	Light/medium
% alcohol	11–12
Starting S.G.	1.085
Final S.G.	Approx. .990
Acidity	0.70% as citric

Ingredients:

Working yeast starter
2 lb. (1 kilo) sweet and
1 lb. (½ kilo) green gooseberries
8 ozs. (¼ kilo) sultanas
½ lb. (¼ kilo) bananas

1½ gm. sodium metabisulphite
3 gms. ammonium sulphate
Pectic enzyme

Sugar and acid to adjust

If no estimation of acidty is made use 6 gms of Citric acid.

Method:

Prepare the ingredients by the method preferred. Add pectic enzyme, ammonium sulphate and 1 gm. sodium metabisulphite. Top up fermentation vessel to almost 1 gallon. Leave for six hours and then take specific gravity and acidity tests. Adjust. Add yeast starter and ferment on "pulp" for approximately three days. Strain into fermentation jar, top up with cool boiled water and fit air-lock. Let fermentation proceed to dryness. Refrigerate for three days. Rack and add ½ gm. sodium metabisulphite. Rack again and fine, as necessary. Store to mature.

Grape

These have been mentioned previously in this chapter and are eminently suitable for all styles of wine if the right varieties are used. The small sultana grapes when plentiful at the greengrocers can be very useful for addition to "musts" containing other ingredients.

Preparation is the same as for other soft fruit but, to obtain a red wine, it is necessary to ferment on the skins of red grapes for two to three days to extract the colour. Even if pure grape juice is used, it is necessary to check the juice for acidity and sugar corrections for the wine style desired, as with other fruit.

Grape: *Wine style, Rose table*

Aim:

Flavour	Light
Bouquet	Light/medium
Body	Light
% alcohol	12
Starting S.G.	1.088
Final S.G.	0.990 sweeten to .998/1.000
Acidity	0.75% as citric

Ingredients:

Working yeast starter · 1 gm. sodium metabisulphite
10 lb. (4½ kilos) sultana grapes · Pectic enzyme
2 pints (1 litre) of juice from · 3 gms. ammonium sulphate
any red fruit, fresh or tinned · 1 gm. potassium sorbate
Alternatively ½ pint (¼ litre) · Sugar and acidity to adjust if
red grape concentrate · necessary

If no estimation of acidity is made use 20 gms of Citric acid.

Method:

Proceed in the usual manner for soft fruit, adjusting specific gravity and acidity and fermenting on the "pulp" for approximately three days. Strain into fermentation jar and let fermentation proceed to dryness. Refrigerate for three days. Rack. Sweeten to 0.998/1.000 and add 1 gm. potassium sorbate to stabilise. Rack again and fine, as necessary. Store to mature.

Grapefruit

These are particularly suitable for aperitif style wines due to their piquant bitter flavour but so often in grapefruit wines there is also a turpentine-like flavour. This is due to the fact that the high percentage of terpenes contained in the oil glands in the outer coloured portion of the skin—the flavedo —oxidise very easily. The essential oils in the skin are also toxic to yeasts and will slow down the fermentation. These considerations, plus the fact that the inner white portion of the rind, the albedo tissue, contains a lot of pectin, pectic enzymes and oxidising enzymes, makes it advisable to cut the fruit in halves and express the juice on a squeezer, or to peel the fruit carefully before liquidising or pressing.

Acidity is high and extension of the juice with water is necessary to reduce and correct it.

Grapefruit: *Wine style, Aperitif, medium dry*

Aim:

Flavour	Medium
Body	Light/medium
Bouquet	Medium
Starting S.G.	1.100
Final S.G.	.990, sweetened to 1.000 approx.
% alcohol	15–20% as desired
Acidity	0.70% or less as citric

Ingredients:

Working yeast starter	Pectic enzyme
Approximately 10 grapefruit or ¼ gallon of juice according to acidity of fruit	2 gms. sodium metabisulphite
	3 gms. ammonium sulphate
	1 gm. potassium sorbate
½ lb. (¼ kilo) minced sultanas	Spirit to fortify if desired
1 lb. (½ kilo) bananas	Sugar to adjust

If no estimation of acidity is made use up to 9 gms. of citric acid to adjust the acidity.

Method:

Take five grapefruit and express the juice. Test for acidity to decide how many fruit will be needed to give the correct acidity in a gallon of wine. Process the extra fruit required. Add 1½ gms. sodium metabisulphite and pectic enzyme. If possible, pressure cook the sultanas and bananas in a little water and, when cool, add to the grape juice. Make up the bulk to about ¾ gallon with cool boiled water. Add ammonium sulphate. Stir well and leave in a covered container for about six hours. Strain off a sample of the juice and test for specific gravity. Adjust to 1.100. Test acidity to confirm original calculations and adjust if necessary. Add yeast and ferment on the "pulp" in a covered container for two/three days, stirring night and morning. Strain into fermentation jar, top with cool boiled water if necessary and fit air-lock. Ferment to dryness. Refrigerate for three days. Rack and sweeten to 1.000 or to taste. Add 1 gm. potassium sorbate and ½ gm. sodium metabisulphite. Rack and fine, as necessary. Store to mature. If it is wished to fortify this aperitif style wine, after the first racking, determine the alcohol content as described in chapter 16, sweeten to taste and fortify as described in chapter 6. In this case there will be no necessity to add the potassium sorbate or the additional sodium metabisulphite.

Loganberry

This is a delicious soft fruit but rather expensive to purchase. Please see reference under "Raspberries."

Lemon

Lemon juice oxidises more easily than any other citrus fruit juice with consequent deterioration of flavour. Lemons are lower in sugar and higher in acid (citric) than oranges and grapefruit and are also high in pectin. They are best used as an additive rather than made as a single fruit wine and the juice should simply be squeezed out and sodium metabisulphite added immediately. This is one case where the unripe fruit is not so acid as the ripe fruit—a reversal of the usual sequence.

Mandarin

These contain citric acid and sucrose. Prepare as lemons. The juice will need extension with water to correct the acidity.

Orange

Here again, as in grapefruit, the carpellary membranes, pith and albedo tissue contain a bitter constituent, as do the seeds. Pectic enzymes and pectic substances are present largely in the inner peel. Oxidising enzymes are also present in the peel.

Citric is the main acid but the content of ascorbic acid is high and soluble nitrogen is also present in the juice.

Orange juice is highly fermentable and the Seville variety have the piquant bitterness suitable for aperitifs. Preparation is by cutting in half and squeezing, metabisulphite being added immediately. As in other citrus fruits, the juice needs extension with water to balance the acidity.

Orange, lemon and honey: *Wine style, Aperitif, medium dry*

Aim:

Flavour	Light/medium
Body	Light
Bouquet	Medium
Starting S.G.	1.100
Finishing S.G.	.990
	Sweeten to 1.000 or to taste
% alcohol	15–20%
Acidity	0.70% as citric
	(or less if desired)

Ingredients:

Working yeast starter	2 gms. sodium metabisulphite
Approx. 1 pint orange juice	3 gms. ammonium sulphate
Juice of 3 lemons	1 gm. potassium sorbate
2 lb. (1 kilo) honey	Sugar to adjust

If no estimation of acidity is made use up to 100 gms of Citric acid to adjust the acidity.

Method:

Express about ½ pint juice and test for acidity. Add juice of three lemons to sample juice and test again. In this way, calculate how many more oranges may be added to obtain the level of acidity required. Process the extra fruit required. Add 1½ gms. sodium metabisulphite, pectic enzyme and 4 pints cool boiled water. Melt the honey in hot water and when cool, add to juice. Add 3 gms. ammonium sulphate.

Take sample and test for specific gravity, adjust to 1.100. Confirm acidity is correct as calculated. Adjust if necessary. Add yeast starter and when the juice is fermenting well, top up gallon jar with cool boiled water and fit air-lock. Ferment to dryness. Refrigerate for three days. Rack and sweeten to 1.000 or to taste. Add 1 gm. potassium sorbate and ½ gm. sodium metabisulphite. Rack and fine, as necessary. Store to mature. Here again, if it is wished to fortify this aperitif, after the first racking, determine the alcohol content, sweeten to taste and fortify. Do not add potassium sorbate or sodium metabisulphite.

Pear

Pears do not usually make good wines on their own, having an insipid taste and need to be blended with other fruits in "musts," apples being suitable and usually available at the same time. About one-third to half as many apples to pears should be used. The skins of pears are high in tannin.

A press or juice extractor is necessary to obtain an economic yield of juice. If a purée is produced rather than a clear juice, sulphite and stand for 24 hours to enable the solids to fall to the bottom of the container then decant off the nearly clear juice. This may be fermented without dilution if the acidity is correct for the style of wine envisaged. Sultanas and bananas are suitable additives if more body is required.

Peach

Different varieties of peaches vary considerably in flavour so it is useful to note the varieties used in the recipe book, for future reference. Sound fruit only should be used. Stone the fruit and liquidise or "pulp" in about 1½ times the volume of cool boiled water. The juice may be strained off after steeping or "pulp" fermentation carried out for two days.

Incidentally, peaches and apricots often have an unpleasant bouquet when the wine is young. However, experiments have shown that if the fruit is pressure cooked for a minute or two at 15 lbs. pressure in a little water, this unpleasant aroma is not apparent in the wine. Perhaps the reader would like to try two batches made by the different methods in order to compare flavour and aroma.

Peach: *Wine style, Table dry*

Aim:

Flavour	Light
Body	Light
Bouquet	Light/medium
% alcohol	12
Starting S.G.	1.088
Final S.G.	.990 approx.
Acidity	0.70% as citric

Ingredients:

Working yeast starter	2 gms. sodium metabisulphite
2 lb. (1 kilo) fresh sliced peaches	3 gms. ammonium sulphate
¾ lb. (350 gms.) sultanas	Pectic enzyme
½ lb. (¼ kilo) bananas	Sugar and acid to adjust

If no estimation of acidity is made use 32 gms of Citric acid.

Method:

Process the peaches by the method preferred, also the sultanas and bananas and place all ingredients in a container. Add 1½ gms. sodium metabisulphite and pectic enzyme. Also ammonium sulphate. Make up bulk to about ¾ gallon with cool boiled water. Cover and leave for about six hours. Strain off sample of the juice and test for specific gravity and acidity and adjust as required. Add working starter and ferment on the "pulp" for two days. Strain into fermentation jar, top with cool boiled water if necessary and fit air-lock. Ferment to dryness. Refrigerate for three days. Rack. Add ½ gm. sodium metabisulphite. Rack and fine, as necessary. Store to mature.

Peach

Method: Wine style, Dessert

To make a dessert peach wine, try 1 pint white grape concentrate, 2 lbs. bananas and as many peaches added as the acidity will allow. Proceed as for other dessert styles given in this chapter.

157

Pineapple

The main flavour of pineapples is to be found in the colloidal constituents of the fruit which is the reason why canned pineapple juice is not clear, as filtering removes the colloids and consequently the flavour. Pineapples used alone, therefore, are not ideal for winemaking. Fresh pineapples may be prepared as for hard fruit whilst tins of crushed pineapple are easy to use for "pulp" fermentation for two/three days. About 1 gm. grape tannin should be added to the "must," if the other ingredients do not contain this.

Plum

If ripe plums are left in a bowl for a few days a wine-like smell will soon become apparent as the fruit begins to ferment with the wild yeasts on the skins. It is therefore necessary to use sound, clean fruit and to sterilise thoroughly before "pulp" fermentation and straining after two/three days. Some difficulty is sometimes found in clearing plum wine, due to the waxy substances on the skins also the flavour and depth of colour will depend upon the variety used. It is best to halve and stone the fruit during preparation when any insect attack inside the fruit will be easily apparent. The quantity of plums per gallon will depend upon the wine style required and the acid/flavour balance.

Pomegranate

This fruit contains a very high tannin content in the skins and pith and the best method of preparation is to halve and squeeze out the juice on a hand squeezer. Claret type table wines have been made in America during experiments with this fruit but it is one of the less popular ingredients for home winemakers in the British Isles, being only available for a few weeks each year, fairly expensive and somewhat messy in preparation. Citric is the main acid. The juice alone may be fermented, with no water extension but, of course, with the necessary acidity and specific gravity corrections and the usual additives.

Raspberry

A delicious soft fruit, which is easily prepared by liquid-ising and steeping or fermentation for one or two days on the "pulp," producing a wine generally with the unmistakable aroma and flavour of the fruit. Loganberries are very similar in composition and style and both are eminently suitable for table dessert wine styles in particular. Due to the high flavour and acidity, care needs to be taken not to exceed the quantity of fruit required. Test the acidity and flavour dilution with a small quantity of fruit before making up a gallon "must" of the style required.

Rhubarb

Sixty to seventy-five per cent of the weight of rhubarb consists of juice which is high in oxalic acid and sharp tasting. The oxalic acid is difficult to remove if the juice is heated.

Many recipes advise removal of the acid with precipitated chalk but often a very insipid wine results, even with the addition of other acid afterwards. Chapter 3 recommends extension of an acid juice with water to the acidity required for the proposed wine style and this is the best method for processing rhubarb.

The fruit should be washed and chopped into small pieces and either put through a juice extractor, or left soaking in layers of measured sugar, pectic enzyme and metabisulphite, until the juice runs free. A word of warning! Do not be tempted to feed whole sticks of rhubarb into the juice extractor, as the long "strings" of the rhubarb will get entangled around the pivot of the spinner.

The Strawberry variety is richer in flavour than the Victoria and the varieties Ruby and MacDonalds Crimson have highly coloured juice.

Rhubarb produces an indifferent wine if made on its own but is very useful as a basic wine to blend with other juices or flavourings, or to extend another fruit base. An example of this is a sound but indifferent rhubarb wine to which a few drops of "Spirits of Orange" have been added. This has the affect of changing the characterless wine into a delightful "orange" wine as regards flavour and bouquet. A drachm or two of Spirits of Orange may be obtained from the local chemist for a shilling or two. "Cheating" or "art"?—this addition makes all the difference!

Strawberry

Like raspberries, this fruit usually produces a wine which smells and tastes unmistakably of the fruit and is, therefore, most suitable for a table dessert style. Preparation is as for other soft fruit, preferably by liquidising. Here again, it is best to process a small quantity of the fruit first to assess the amount of dilution required to produce the desired flavour and then to adjust the acidity. Critically assess the flavour on the palate without sweetening the juice.

Tangerine

Cut in half and extract the juice on a hand squeezer. The juice will need to be extended with water, as for other citrus fruits. A medium sweet wine will need approximately 12 tangerines per gallon with ingredients decreased or increased proportionately for other wine styles.

Dried fruit

Dates, currants, figs, raisins, sultanas, prunes, elder-berries and bilberries are best cooked in about ½ pint water to each 1 lb. fruit for 10–15 minutes pressure at 15 lbs. Bananas, apricots and peaches, rose hip shells and rose hips need five/six minutes pressure cooking at 15 lbs. pressure with about ½ pint water for each 1 lb. of fruit.

Pressure cooking will completely sterilise the fruit, which in all probability has a high micro-organism population due to storage and the fruit will then be soft and ready for extension

160

with additional cool boiled water and subsequent maceration and "pulp" fermentation.

However, if a pressure cooker is not available, to every 1 lb. of fruit, heat 4 pints of water. Wash and clean the fruit, shred or mince where practicable and then soak in the boiling water in a sterilised vessel. Stopper with a cotton wool plug moistened with sodium metabisulphite and, when cool, add 1½ gms. sodium metabisulphite for each gallon of juice, as explained in chapter 3.

Do remember to take into account the likely colour extraction when using dried fruit to blend with a main ingredient. For example, if making a white tablewine, sultanas are preferable to raisins, being much lighter in colour, in order to obtain a light straw colour wine rather than a brown colour. Similarly, a red table wine needs to be a "black red" and not a "brown red" and therefore care must be taken in the choice of additives. Generally speaking, the dried fruit which produces a brownish colour should be reserved for dessert styles and not table wines.

Rose hip and fig: *Wine style, Aperitif*
Aim:

Flavour	Light
Body	Light/medium
Bouquet	Light/medium
% alcohol	15–20
Starting S.G.	1.100
Final S.G.	.990 sweeten to taste
Acidity	0.50% as citric

Ingredients:

Working yeast starter (Sherry if possible)
½ lb. (¼ kilo) rose hip shells
2 ozs. (50 gms) figs
2 lb. (1 kilo) fresh bananas
1 pint (½ litre) white grape concentrate, if desired

2 ozs. (50 gms) malt
1 gm. tannin
Pectic enzyme
3 gms. ammonium sulphate
2 gms sodium metabisulphite

If no estimation of acidity is made use 20 gms of Citric acid.

Method:
Pressure cook the rose hip shells, figs and bananas. Place in sterile container with the grape concentrate and malt and

blend together in the hot liquid. When cool, add 1½ gms. sodium metabisulphite, pectic enzyme and tannin, also ammonium sulphate. Extend with cool boiled water to ¾ gallon. Cover and leave for about six hours. Take sample of liquid and test for specific gravity and acidity and adjust. Add working yeast starter and ferment for three/four days on the ingredients. Strain into fermentation jar, top up with water if necessary and fit air-lock. Ferment to dryness. Refrigerate for three days and rack. If it is desired to sweeten the wine, at this stage sweeten to taste and add 1 gm. potassium sorbate. If it is required to emulate a sherry style wine do not add the additional ½ gm. sodium metabisulphite which is usual in a dry white wine, as slight oxidation is a characteristic of this style. The reader may wish to fortify the wine, in which case, determine the alcohol after the first racking, sweeten if required, and fortify. Refrigerate for one week and rack again. Rack and fine, as necessary. Store to mature.

Grape concentrate and honey

Wine and mead respectively can be made from these ingredients used individually but they are also useful added to other ingredients to make quality wines and to add body. If it is desired to ferment honey on its own to make mead, one with little flavour is preferable, clover honey generally being suitable. Both ingredients need to be melted in hot water and, when cool, 1½ gms. sodium metabisulphite added to prevent bacterial infection.

Other concentrates

For quick and easy winemaking, other concentrates are also available, such as orange, rose hip syrup, apple, ginger, etc.

It will be found that a judge will leave a ginger wine to be tasted last if such a wine is found in a competition, as the ginger flavour will linger on the palate but for those lovers of ginger wine, a recipe is given below.

Ginger: *Wine still, Dessert, unfortified*

Aim:

Flavour	Full
Body	Full
Bouquet	Full
% alcohol	15
Starting S.G.	1.100
Final S.G.	.990 sweeten to approx. 1.020
Acidity	0.50% as citric

Ingredients:

Working yeast starter	Pectic enzyme
2 pints (1 litre) ginger concentrate	2 gms. sodium metabisulphite
1 lb. (½ kilo) sultanas	3 gms. ammonium sulphate
2 ozs. (50 gms.) figs	
Sugar and acid to adjust	1 gm. potassium sorbate

If no estimation of acidity is made use 19 gms of Citric acid.

Method:

Melt the concentrate in hot water. Pressure cook the sultanas and figs and add to the concentrate. Top to about ¾ gallon with cool boiled water and when cool add pectic enzyme, 1½ gms. sodium metabisulphite and 3 gms. ammonium sulphate. Take sample of liquid and test specific gravity and acidity and adjust. Ferment on "pulp" for three days approximately and then strain into fermentation vessel. Top with cool boiled water if necessary and fit air-lock. Ferment to dryness. Refrigerate for three days. Rack. Sweeten to approximately 1.020 or to taste. Add 1 gm. potassium sorbate and ½ gm. sodium metabisulphite. Rack and fine, as necessary. Store to mature.

Flowers

Fresh flowers should be picked on a dry day and left in a container open to the air as if they are enclosed in a plastic container, the delight of the aroma often turns to disgust.

163

This particularly applies to elderflowers as these are notorious for the tendency to give wines a "catty" aroma but they can also give a delightful bouquet. It has been mentioned previously that there are many varieties of these bushes, so it is important to smell the flowers before picking them. If the aroma is pleasant, take them home in an open container.

Wash the flowers, enclose in a muslin or net bag and immerse in the fermenting liquid until the required aroma and flavour has been extracted, when the bag may be removed. Nose and taste tests can be carried out, say, night and morning and, in this way, the aroma and flavour can be regulated to the maker's preference.

Dried flowers may be immersed and treated in the same way. Sometimes this infusion may be carried out at the end of fermentation if a wine is lacking in bouquet.

An alternative method of using flowers is to make a highly concentrated flower wine on a fruit base with the sole idea of adding small amounts to other wines to add bouquet when required.

CHAPTER 12

Tasting and Glossary of terms

THE systematic way of making a wine without slavishly following a recipe and how to store it and serve it at table have been dealt with in previous chapters—but what of the all important taste assessment of the wine?

This book is written with the intention of providing anyone who makes wine in the home with a concise coverage of the various technical problems likely to be encountered and to explain them in terms everyone can understand. Critical tasting, for too long the preserve of "experts," has until recent years been surrounded by mystic and assorted terminology, commercial and amateur tasters being equally guilty. The Glossary of Terms and descriptions given at the end of this chapter is an attempt to provide a degree of standardisation for amateurs so that when they compare notes the terms will have meaning to everyone in the discussion and not only the person who made them. A classic example of keeping accurate notes which no one could understand for over a hundred years, was the diary of Samuel Pepys; surely today no reason exists for writing in jargon others will not understand.

Tasting Hints

For critical tasting the best time is about 11 a.m. By this time the after-taste of breakfast and the effects of the night before will have worn off and the palate and is at its most receptive. After lunch the meal will have dulled the most acute faculties of the palate and any assessment made is unlikely to have the clarity of one made in the morning.

Scientific evidence shows that smoking is not such a disastrous thing as far as the palate is concerned as many people think and probably the best advice that can be given is, if you smoke and taste wine, keep a careful note of your observations and then the next day, or at the weekend, repeat the tasting but this time *not* smoking. If, when you compare the notes of the two tastings, your senses were sharpened by not smoking, the answer is obviously that smoking affects your palate. If, however, little difference exists, it is also obvious that your palate is attuned to smoking and makes little difference to your tasting ability. How much better the palate might have been if you had never smoked, is an unanswerable question. Needless to say, consideration has to be given when tasting with non-smokers, cigar and pipe smoke being especially objectionable to non-smokers at these times.

The only equipment required for tasting is a sheet of white paper, one or more glasses of the tulip or tasting type shown in chapter 10, a table lamp with shade removed, a notebook and a pencil. An indexed book or card index will make it easier to find notes when comparisons need to be made. At one time a candle was considered the only light by which to judge wine but this is really useful only for assessment of clarity.

Smell the glass intended for tasting, as a glass cloth can impart a quite undesirable aroma to a glass during wiping. For this reason, glasses are often washed, rinsed in cold water and left in an inverted position to drain dry.

Rule up the book or cards you are going to use for your tasting notes with five columns. The first two need be no more than 1″ wide but the other three need to be wider. Head the columns from left to right: Clarity, Colour, Nose, Palate and General Remarks.

The order in which to examine a wine is clarity, colour, bouquet and, finally, palate. Get into the habit of writing down your observations whilst actually looking, smelling or tasting the wine as the case may be—notes are much more accurate made in this way. When more than one style of wine is to be tasted at a session, it is best to taste white before red and dry before sweet to preserve the sensitivity of the palate.

Critical tasting is invaluable in gaining experience and knowledge to improve the standard of one's own wine—and one should get into the habit of doing this quietly and seriously for this reason alone—apart from encouraging the ability to assess other amateur wines and commercial products. Properly done, bearing in mind some of the following points, a great deal can be learned from these observations.

Taking a taste sample

Write down first the name of the wine and its age.

Clarity

Before pouring out any wine from the bottle, note if a sediment or deposit exists. If it does, make a note, also indicating if it is fine, coarse or crystalline and if it is coloured. A brown deposit in a white wine would indicate the wine was contaminated with copper or severe oxidation had taken place, whilst a white or creamy breadcrumb-like deposit is more likely to be yeast which has carried out a secondary fermentation after the wine was bottled. A red wine will often throw a deposit consisting of tannin, colouring and/or protein matter: nevertheless it should be noted as it may give a guide to the "bottle age" of the wine or explain some observation made later. Brown wines can throw a brownish deposit of unstable protein matter, which is due to the inclusion of colouring in the deposit. Having carried out the preliminary examination, carefully pour (or decant) enough wine to fill the glass, which must be free from taint, about one-third full.

Using the electric light, examine the wine for clarity. All wines should be clear. If any fibres are present, make a note to this effect. They could have come from the filter.

Colour

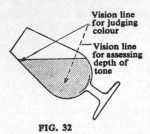

Vision line for judging colour

Vision line for assessing depth of tone

FIG. 32

Tip the glass at an angle of 45° over the white paper and examine the "rim" as shown in fig. 32.

By using the Glossary of Terms, note the colour of the wine. White is meaningless except as a general classification and can range from water-white to amber when used in connection with wines so something more explicit is required. Explicit notes are also necessary with red, rosé and dessert wines—describing the depth of colour takes a little practice and the Glossary explains the meaning of "full," "medium" and "pale."

Nose (or bouquet)

A great deal can be told from this assessment. The olfactory senses are able to distinguish the aroma of the ingredients—such as the odour of raspberries in a young raspberry wine—as well as the esters apparent in a mature wine—and these characteristics form the bouquet.

Besides assessing the quality of a wine, another important reason for smelling it is for guidance as to whether it should be tasted or not. If a smell like nail varnish, or wine vinegar, cork, rain-water/butt or any other offensive smell is apparent, then of course the wine should not be tasted, or at least not until any other tasting has been done.

If a wine has little or no bouquet, this could be because it is very young or lacking character and until the tasting is complete, a conclusion cannot be drawn. Young wines have a yeasty or rather aldehydic character which is sharp on the nose whilst wines which have aged to some degree are softer and do not have this sharpness.

An assessment of the depth of bouquet, if a fruity aroma is present and if so, how much, and the general impression of the aroma should be noted.

Palate

Many of the opinions formed from a study of clarity, colour and bouquet are crystallised into a complete assessment by this appraisal.

The things to be looked for are:

1. **Body:** This means the feel of the wine in the mouth, if it is light and rather delicate or heavy and robust. This is, amongst other things, an indication of the alcohol and extractable solids present. A pronounced taste of alcohol indicates either the wine was more sugar and water than anything else or that, in the case of a fortified dessert wine, the fortification was relatively recent and the wine has not "married" together with the spirit properly.

2. **Acidity:** Some acidity is necessary to provide balance and keeping properties but it should not predominate.

3. **Sweetness or Dryness:** This is self explanatory to a large extent but what needs to be understood is that astringency—a drying of the roof of the mouth and clinging to the teeth—is not the same as a dry finish due to all the sugar being fermented into alcohol—one is due to tannin and the other to sugar, or lack of it.

4. **Flavour:** The important thing is to record what is tasted in terms that can be understood by someone else. If the flavour is reminiscent of blackcurrrant juice, then this is the best way to describe it. Fruitiness is a fleshy texture more than a flavour and shows the fruit origin rather than just a flavouring.

5. **Balance:** When a wine is balanced, all characteristics harmonise, with nothing predominating in an offending manner. The body, alcoholic content, flavour, tannin and acidity combine to form the wine style required. For instance, a sweet wine should not in any way appear dry either in the taste of after taste, otherwise it would be unbalanced.

A specimen record card or sheet is shown on pages 170-1. Using a record of this nature most of the facts relevant to making a particular wine can be found easily together with an assessment of its taste in meaningful terms.

FRONT

WINE RECORD CARD

Wine: *Red-currant*

Notes: *A dry red wine of about 12.5% was required.*
2 lbs. of fruit was crushed and mixed with 4 pints of water. 1 gm. of sodium metabisulphite was then added. After 6 hours a sample was strained off and the S.G. and acidity determined. An acidity of 0.55% as citric acid was aimed for.

Initial volume of juice: + *Water 5½ pints*

Specific gravity: *1.038*

Total acidity: 0.25% as citric acid
After sugar and volume made up to 2 gallons.

Additions

Sugar: 2 lbs. 10 ozs.   **Acid:** 1 ozs.

Final S.G.: *1.088*   Final acidity 0.56% as citric acid

Final volume: 2 gallons

Other additions: Sodium metabisulphite: 2 gms. in all   **Pectic enzyme:** about ½ gm.

Fermentation

Started: *20/7/1973*   Finished: *31/7/1973*

Date	20th	21st	22nd	24th	25th	26th	27th	28th	29th	1st/8		
Temp. °F.	67	72	72	74	74	73	73	70	70	71		
S.G.	1.085	1.060	1.040	1.015	1.013	1.008	1.005	1.002	1.000	1.000		

Clarification

Racked: *3/8/1973*   Fined/Filtered: *5/12/1973*   Fining used: *Bentonite*

Bottled: *1/1/1974*

170

BACK

TASTING

Clarity	Colour	Nose	Palate	Remarks
Tasted 20/2/1975 Clear and bright.	Medium tawny	Clean and fruity with a blackberry-like aroma.	Fruity but not as much as expected from the nose. A little alcoholic but otherwise balanced. Medium sweet and mellow	Mature ready for drinking. The slightly disappointing palate compared with the nose could be due to slight overaging.

Glossary of Terms

1. Clarity

Brilliant   That degree of clarity which makes the wine "sparkle" like cut glass.

Clear   Indicates that no deposit is present or haze suspended in the wine and that filter fibres or any other foreign bodies are absent.

Clot   A clot on the top of the wine indicates very inefficient pectin removal. Small clots in the wine also indicate pectin or unstable protein matter.

Deposit   A sediment formed at the bottom of the bottle. Secondary fermentation produces a yeast deposit which is off-white and can be either fine or crumb-like, depending on the yeast. Metal contamination produces a deposit or casse the colour of which is usually off-white or grey in the case of iron whilst copper is commonly brown.

Hazy   A fine haze or cloud. This could be the result of poor fining removal, metal contamination or protein deposits.

Homogeneity   Any sign of mixing should be suspected as it suggests poor blending just before bottling.

Silky haze   A silky sheen on the haze. This usually indicates a bacterial infection. Often the bottle has to be swirled to see this effect.

2. Sparkling   A good gas content in the wine. It effervesces well and does not quickly go flat.

3. Colour   The term "red" is meaningless without qualification, as any motor car dealer knows. The same is true in relation to the colour of wines.

(a) Tint

Colourless   Water-white—as in Vodka or Gin.

White   Means very little on its own and should be avoided except to classify broadly.

Pale straw   A colour like that of a drinking straw.

Straw   The colour of a straw stook as seen on a farm. This is a pleasing pale yellow colour.

Pale lemon   The colour of clear pure lemon juice, freshly pressed with just the faintest tinge of green present.

Golden   A bright golden hue as seen in sugar syrup. This colour is often associated with sweeter wines.

Brown/gold   The appearance of a brown tinge in a "white" wine usually means oxidation has set in and the wine is past its best. In a dessert wine it is quite normal or where dried fruit has been used which is rich in caramelised sugar.

Pale brown   The colour of an amontillado sherry. Quite a good shade for an aperitif.

Mid-brown   The tint of mild ale. A good shade for a dessert style wine.

Brown   A shade like chop sauce. Not a very good colour except for the heaviest dessert wines.

Rosé   The delicate pink seen in rose petals and should be reserved for describing the light tinge of true rosé wines.

Pink   Self-descriptive but should not have any orange or blue tinges. Often used for pink wines of deeper hue than rosé.

Orange/pink An orange tinted pink. This can be attractive and yet still not be pink or rosé. Sometimes it indicates the blending of white and red wines.

Orange	A pale orange tint usually indicates the blending of wines to produce what is hoped to be a rosé.
Ruby	A red colour like that of Port Wine.
Red-brown	The pure red changing to a shade which has a little brown in it. This indicates the wine has some maturity unless a blending of red and brown wine has taken place.
Purple	A blue red colour. Young wines often have this colour.
Mahogany	A rich deep tone of genuine mahogany furniture. When this is noted in what was a *red* wine either the wine has matured very well or has deteriorated through oxidation.
Tawny	A light mahogany shade.

(b) Depth of tone (important for red wines)

Pale	The depth of colour is not very great. For example, a bottle of blackcurrant syrup has great depth of colour but a teaspoonful of syrup in a glass of water will have the same tint but be pale. Thus a wine could be described as "Pale Red/Purple."
Medium	The next step up from pale. Most ruby Ports or Beaujolais wines could have this description.
Full	The depth of tone encountered in a good elderberry wine or, in the wines of commerce, a Chateauneuf du Pape. Almost, but not quite, translucent.

NOSE (Bouquet)

The important thing here is to write what is smelt in terms you and others will understand.

Aroma	Smell
Asbestos	A smell like that produced when boiling water is poured on to asbestos "pulp." The filter was incompletely washed.

174

Burnt	Either excessive cooking of ingredients or storage too near a heat source.
Caramelised	This smell is a sign that the wine has either been made from fruit which was over-cooked or it has been allowed to overage and become oxidised.
Clean	No "off" flavours present.
Corky or musty	The smell similar to a dry cork which has been in a moist but otherwise empty medicine bottle for some time.
Earthy	A smell akin to damp soil. If Bentonite fining has been done it might mean too high a dose was used.
Fragrant	A perfumed aroma, delicate and flowery.
Fruity	Indicates the aroma is like that of a fruit and is best qualified with a description of which fruit, viz., "Clean and fruity with a banana-like aroma."
Green	A term often used to mean young and immature; it is characterised by an aldehydic aroma and often a yeasty smell.
Peardrops or a nail varnish like smell	A danger sign, often an indication that the wine is turning slightly acetic. A bacterial infection or oxidation are immediately suspected.
Rotten eggs	Wines which have just finished fermenting sometimes have this aroma due to hydrogen sulphide gas produced by the yeasts reducing sulphur bodies dissolved in the wine. This will gradually disappear with racking and subsequent operations to clarify.
Faint Medium Full	Used to describe the depth of aroma. Faint means little bouquet at all and so on.

Spirity	Alcoholic aroma can mean that the wine is rather raw and unbalanced or that fortification has taken place.
Stalky	The smell of freshly cut twigs. Often indicates a young wine.
Sulphury	The smell of a London fog or air near a brickworks. A sign of too much sulphur dioxide.
Sweet	Self-explanatory.
Woody	A smell like a rain water butt which indicates the wine was at sometime in a poor wooden cask or was left too long in wood.
Yeasty	A smell like bakers yeast often present in new young wines which will gradually disappear with racking and ageing.

PALATE (Taste)

Alcoholic	The pronounced flavour of alcohol indicates very little natural ingredients (sugar and water used too well) or poor fortification.
Acid	A wine must have some acidity to give it zest and prevent it being flat or insipid but it should not predominate so much as to give tartness and overpower all the other characteristics.
Astringent	A drying of the mouth. When due to excess tannin, this is unpleasant. A little tannin in a red wine is, however, desirable.
Balanced	A satisfactory combination of alcohol content, acidity, tannin and flavour to give a pleasing taste.
Bitter	A taste sometimes encountered due to wrong fruit handling. Some fruit skins if fermented with the wine will give this taste. See also Metallic.

176

Body	The feel of the wine in the mouth—its weight due to alcohol and extracted solids present. It can be light, medium or heavy.
Cooked	A caramel like flavour encountered when the wine ingredients were either cooked or over-pressure cooked or the temperature of the fermentation was rather high.
Coarse	A roughness associated with rather poor wine.
Cloying	Sweet and heavy not offset with the necessary acidity.
Dry	All the sugar fermented out. This does not mean the mouth is dried out or astringent.
Earthy	Either poorly chosen or incompletely cleaned ingredients or overfining with Bentonite.
Fresh	Pleasant fresh acidity and refreshing to drink.
Fruity	Taste of fruit. If possible an idea of what fruit the taste is reminiscent should be recorded. It does not matter if this was not the fruit used—conveying what was tasted is important.
Flat	Lacking zest. Can be due to low acidity or alcohol or over-ageing.
Green	Young and rather immature. Indicates a lack of "marrying" together which will come with age.
Hard	Severe taste due to excess of tannin.
Harsh	Alcohol predominates in the taste over everything else.
Insipid	A wine lacking character, flat and without the necessary acidity.
Mellow	Smoothness due to the glycerol of the wine and age.

Metallic	A bitter clinging taste due to contamination of the wine with probably iron (less often from copper).
Medium dry	A basically dry wine with just a trace of sweetness.
Medium sweet	A wine which is sweet but not in a pronounced manner.
Mousey	A most unpleasant and often bitter taste and after-taste accompanied by the smell of mice can indicate a bacterial infection. A haze should be carefully looked for in an attempt to support this.
Prickle	A tickle on the tongue due to dissolved carbon dioxide in the wine. In what is supposed to be a still wine, this indicates a secondary or malolactic fermentation has taken place whilst the wine was in bottle.
Robust	A term used for red wines to mean balanced, full-bodied, and full-flavoured with a fruity texture.
Round	All the characteristics of a wine integrated to form a mellow, smooth quality.
Rich	A wine having qualities full and luscious to the palate and senses.
Smooth	No obvious excess of acidity or tannin—a well-balanced and mature wine.
Soft	Bland, astringency absent and agreeable to the palate.
Sweet	Residual sugar in the wine.
Tannin	Tannin is evidenced by a drying of the mouth and a grip on the teeth. It can mean a very young wine is being tasted which will in the course of aging precipitate much of the excess—or a poorly made wine with too much tannin in the fruit or added tannin.

Tart	A wine giving the impression of unripe fruit or excess acidity.
Woody	A twiggy taste which can be due to the wine being stored in a cask with a new immatured stave.
Unbalanced	A wine with a lack or excess of acid, fruit, tannin or body and where the constituents do not harmonise and balance each other to form the wine style desired.

CHAPTER 13

Micro-organisms of fermentation and spoilage

THE micro-organisms which concern the amateur winemaker all belong to the plant kingdom and form part of the sub-kingdom called the Thallophyta. These plants do not flower, have leaves or form wood and are much simpler in botanical structure than the flowering plants and, because many of their features need to be studied with the aid of a microscope, the science involving these organisms is called microbiology. Broadly speaking these organisms can be divided into three classes as follows. None of them have chlorophyll and therefore they cannot synthesise food from inorganic compounds with the help of sunlight; all their food comes from the organic matter upon which they live.

Bacteria

These consist of single cells which are either rod shaped or spherical and multiply by a splitting of the cell down the middle to form two new cells. They are very small in size and can only be seen by a microscope and even this, unless it is very high powered, will show very little of their structure. Bacteria are responsible for the transformation of malic acid to lactic acid in the so-called malolactic fermentation and also for the acetification of wine which is left exposed to air. Another spoilage condition for which bacteria are responsible is ropiness, a condition fortunately not very common in amateur wines.

One important fact, however, is although the bacteria found in association with wine will cause spoilage, they are *not* harmful to human beings by causing disease.

Yeasts

These, once again, are usually single cells and do not form mycellium—chains of cells each joined to the next—which is a characteristic of moulds. Sometimes yeasts form chains of cells but they do not unite together into a true mycellium.

Yeast cells reproduce by budding and/or by the formation of spores and depending on which path they follow so they are classified. This will be explained in more detail later.

As already explained, yeast cells contain enzymes which are responsible for alcoholic fermentation and, of course, it follows that as the number of yeast cells increases so does the concentration of enzymes until a point is reached where the yeast cells are killed by the alcohol content of the fermentation which also inactivates the enzymes.

Moulds

The characteristic feature of moulds is the production of cottonlike white threads called mycellium. When seen under the microscope some moulds have mycellium with cross walls and are called "septate" whilst others do not have this and are referred to as "non-septate."

The moulds of interest to the amateur winemaker can be differentiated roughly by their colour, penicillium being green, Aspergillus being black, Mucor being whitish and brown, and Rot being brown.

The moulds reproduce by forming spores which in many cases are air borne and contamination of fruit especially is thus very easy.

Having broadly classified the organisms, we can now look at the specific ones which may be encountered, see what symptoms will be evident and what action to take. For this it is more convenient to look first at the moulds, then the yeasts, and finally bacteria because it is in this order the organisms are likely to be found in practice.

Occurrence and control

Moulds

These usually require a free supply of air for their growth but some species exist which can grow in bottled fruit juices, for instance, where sufficient air has remained in the bottle after filling even though the juice has been filtered and the bottles sterilised. The reason for this is that mould spores are extremely small and can pass through a filter. To prevent this happening, if juice is to be stored, the bottles should be *loose* stoppered and placed in a pressure cooker for 15 minutes at 10–15 lbs. per sq. inch pressure and then screwed down firmly on removal. In this way air will be expelled from the pack and the juice sterilised. Some slight change in the flavour of the juice occurs as a result of this process but it is far preferable to a mould growth. Juices should not be oven sterilised as the heating is too severe and caramelisation occurs. These observations are of interest mainly to those who wish to keep fruit juice for a few months and then ferment it when fresh fruit is not available.

Sometimes one hears of a juice growing a mould before fermentation has started. Mainly this is due to poor cleaning and sterilisation, but not always. Spores can easily settle on the surface of freshly racked juice and if the fermentation is slow to start the mould will grow in the presence of air. The remedy is to get the fermentation off to a good start in the first 24 hours; in this way an atmosphere of carbon dioxide will be produced above the fermenting juice and since moulds will not flourish under these conditions, no problem arises.

Mould growths on juices can lead to the inclusion in the juice of minute quantities of mycotoxins. The best course of

action if a mould growth occurs on juice is to dispose of it
and avoid any risk to your health.

Yeasts

1. Fermentation yeast

Seen under the microscope a wine yeast (*Saccharomyces Ellipsoideus*) appears as in fig. 33.

Fig. 33

Some cells are shown constricting the cell wall near one end to produce a daughter cell. This is budding and can easily be seen if a drop of fermenting wine is looked at under even a comparatively low powered microscope. Some of the other cells have within their walls spores or more correctly ascospores which, when released from the enclosing wall, themselves grow into yeast cells. The ability to form these spores is one of the features used to classify yeasts.

Saccharomyces Ellipsoideus is the true wine yeast and has the ability to ferment sugars quickly up to approximately 13% by volume alcohol and more slowly to approximately 15.5%.

Various species of wine yeast exist and are sold as Champagne, Burgundy, Bordeaux, etc. The reason for this is that just as plants and animals adapt to the environment in which they grow, so do yeasts. Changes of this nature involve modifications within the cells and concern the genes or inheritance-carrying factors so that, whilst the profile of the cells appear very much the same under the microscope, the various strains do produce different effects on the fermentation.

2. Apiculate yeast (or wild yeasts)

One of the most prolific yeasts during the early stages of fermentation is a lemon shaped yeast which does not form spores, called *Kloeckera Apiculatus*. This is commonly referred to as Apiculate yeast and is shown in fig. 34.

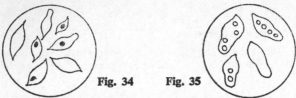

Fig. 34   Fig. 35

Kloeckera Apiculata   Saccharomyces Ludwigii

The black spots are not spores but cell nuclei—or governing centres. Reproduction is by budding. These yeasts do not convert sugar to alcohol as efficiently as *Sacch. Ellipsoideus* and produce greater quantities of aldehydes; their tolerance to alcohol is low, being capable of fermenting only up to about 5.5% by volume alcohol before being killed.

Yeasts belonging to the genera *Saccharomycodes Hanseniaspora* and *Nadsonia* also have lemon shaped cells but do form spores. They also carry out fermentation.

It is undesirable to have apiculate yeasts start a fermentation because of the tendency to produce "off" flavours, fortunately they are susceptible to sulphur dioxide and a dose of 100 ppm. SO_2 added to the juice will ensure a clean start by the cultured yeast selected.

Other yeasts will also carry out fermentation if given the opportunity and among these are yeasts belonging

Fig. 36

Brettanomyces Anomalus

to the genera *Torulopsis*, *Brettanomyces*, *Hansenula* and *Pichia*. If a wine containing residual sugar is neglected or handled in a careless manner allowing contamination to enter either from the atmosphere or from incompletely cleaned and sterilised apparatus any of these organisms or, of course, the wine yeasts may carry out a further fermentation. *Torulopsis Colliculosa* in appearance closely resembles *Sacch. Oviformis*. *Brettanomyces* yeasts have more sausage shaped cells and produce a considerable increase in volatile acidity. The *Hansenula* and *Pichia* yeasts are less common and more likely to infect juice rather than the finished wine.

3. **Film yeasts**

Often yeasts in this category are called "wine flowers" or "flor" yeasts. They have an ability to grow on the surface of wine in a film-like growth after the main alcoholic fermentation has ceased. In the early stages the film appears as a white dust spread over the surface and is easily broken. As it develops the film becomes crinkled and forms a skin over the whole wine surface. These yeasts require a free supply of air and for this reason are called Aerobic.

In this group are *Sacch. Oviformis, Sacch. Bayanus* and *Sacch. Fermentati,* which all produce oxidising effects, giving rise to higher concentrations of aldehyde in the wine than normal. It is these so-called "sherry yeasts" which give the Fino wines of Jerez their particular character. These yeasts are resistant to sulphur dioxide and can be responsible for causing fermentation in up to 20% by volume alcohol and for this reason, if no other, bottling should be carried out with care using clean sterile equipment to prevent secondary fermentation after bottling if any residual sugar is present.

Although similar in effect, each of the three yeasts named above are different in appearance under the microscope. They are shown below in fig. 37.

Fig. 37

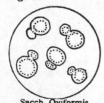 

Sacch. Oviformis   Sacch. Bayanus   Sacch. Fermentati

Although not drawn to scale the variation in size between species is approximately correct.

Some yeasts in the genus *Candida* form a surface growth on wine in the presence of air that attacks the alcohol, extracted solids and sometimes even the organic acids. It is rare for these yeasts to cause alcoholic fermentation. The effect on the wine is to alter the balance adversely. Growth occurs most easily on low alcohol content wines.

Fig. 38

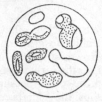

Candida Mycoderma

Many fruit juices and wines provide excellent media for one or more of the yeasts mentioned above to multiply. These organisms exist quite widely in nature and on the skin of fruit; clean, sterile working and the judicious use of sulphur dioxide (by way of metabisulphite) will reduce the risk to a level unlikely to present trouble and since the film yeasts are aerobic care should be taken to keep all vessels containing wine topped right up to exclude air.

Mention has been made several times already of the microscope and its use to see yeasts and other micro-organisms. Obviously the majority of amateur winemakers do not possess such an instrument, as it is not essential for their purposes, but when an opportunity presents itself at an exhibition or the like it is always worth while seeing for one's self what these fascinating microbes look like. In this way a true appreciation of their respective sizes is obtained.

Bacteria

These are so small that millions could rest on the head of a pin; seen under the microscope they appear as little rods (bacillus) or small spheres (cocci) and to attempt anything other than to look at the principal effects they can have on wine would involve the reader in technicalities requiring considerable background knowledge. The following is therefore intended to explain a little of the changes bacteria can make in wine and how they can be detected and prevented from growing.

186

It has already been said that a microscope can be most useful in seeing micro-organisms to forecast what their effect will be but a lot can be learned without this sophisticated aid by using a few simple tests to gauge what is happening.

When a wine is attacked by bacteria a haze is often produced and if the organism attacks the acids present, such as in the malolactic fermentation, an increase in pH and a decrease in total acidity occurs. If the bacteria attack sugars, glycerine or other non-acidic substances, acid is produced as a result of the infection and a decrease in pH and an increase in total acidity occurs. A change in the volatile acidity of the wine often takes place but to measure this requires proper laboratory equipment and is beyond the scope of the amateur.

The attack on acids by bacteria can be swift, a one-third decrease in acidity over a three day period is not uncommon, and if left unchecked will destroy most or all of the acid present.

With bacteria attacking the non-acidic substances the situation is less severe because, with the production of acids, bacterial growth slows down and finally stops. Below pH 3.5 the bacteria which can infect wine find it very difficult to grow and it is for this reason, apart from reasons of taste, that a wine must have sufficient acidity.

Acetobacter; wine to vinegar!

Despite the fact that wine has been made for thousands of years, it was not until just over a century ago, in 1861, that spoilage was clearly established to be associated with micro-organisms. Louis Pasteur, the great French scientist, discovered that some changes in wine acidity were due to microbial changes. This explains why a wine left exposed to air changes to vinegar; a bacteria since called Acetobacter infects the wine and this has the property of being able to use the oxygen present in the air to convert some of the wine alcohol first to acetaldehyde and finally to acetic acid; the chemical reaction is shown overleaf. Acetobacter is an aerobic bacteria, in the early stages a wine infected can show a slightly oily appearance on its surface and have a smell which, although more acrid, does not smell like vinegar or acetic acid. In later stages the wine becomes cloudy, the oily film becomes more obvious and the wine smells of pear drops due to the combining of the acetic acid formed with alcohol in the wine. If allowed to continue the oily film thickens to a scum and a pronounced vinegar smell becomes evident. Because air is essential for the growth of these organisms care should be taken to keep storage vessels topped up and when bottles are filled the air space kept to an absolute minimum.

$$CH_3CH_2OH + O \longrightarrow CH_3CHO + H_2O$$
ethyl alcohol oxygen   acetaldehyde Water

$$CH_3CHO + O \longrightarrow CH_3COOH$$
acetaldehyde Oxygen   acetic acid

Lactobacilli and the malolactic fermentation

Pasteur also found in his research a bacterial infection of wine that produced lactic acid by breaking down tartaric acid, sugars or malic acid. A great deal of work has been done since and it is now known that a long rod bacterium under the general title Lactobacilli is responsible for these changes. These bacteria do not require air to grow and are therefore Anaerobic. Wines badly infected with lactobacilli exhibit a silky sheen when swirled and when poured have an oil-like viscosity; often a smell like that associated with pet mice is noticed. The fig. 39 gives some idea of the difference in size between these bacteria and acetobacter when seen under the microscope × 500.

The presence of residual sugars and nitrogen nutrients favours the growth of these bacteria and for this reason, as well as others already mentioned, wine should not be allowed to stand on the lees after fermentation as this provides good nutritional opportunity.

The discovery that malic acid, which occurs widely in fruits and juices, was changed to lactic acid explained why some wines showed a decrease in acidity and an increase in pH at the same time becoming more smooth. In the formulae following the reader will see that malic acid has two —COOH groups whilst lactic acid has only one. These groups are the ones that give the acid characterisation, thus malic acid is said to be dibasic (two acid groups) whilst lactic acid is monobasic (one acid group). During the malolactic fermentation then, a loss of acidic groups occurs with the formation of carbon dioxide, which vents to the atmosphere if given the opportunity and lactic acid which is a smoother acid than malic is formed. A reduction in acidity and a rounding off thus takes place; for some commercial winemakers this is very important and necessary, for the amateur, without the same degree of control, distinct problems exist. During the malolactic fermentation, because of the reduction in acidity, attack by other bacteria is made easier and it is quite likely that these will include bacteria which will decompose other acids. Citric acid in particular decomposes to form acetic acid, a most undesirable addition to any wine.

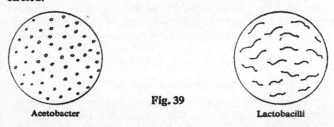

$$
\begin{array}{c}
CO_2 \\
\uparrow \\
\boxed{COO}H \\
| \\
CH_2 \\
| \\
CHOH \\
| \\
COOH
\end{array}
\quad
\xrightarrow[\text{lactobacilli}]{\text{malolactic}}
\quad
\begin{array}{c}
CH_3 \\
| \\
CHOH \\
| \\
COOH
\end{array}
\quad + \quad CO_2
$$

malic acid   lactic acid   Carbon dioxide
(dibasic acid)   (monobasic acid)   gas

Malic acid loses the carbon dioxide from the position circled.

Fig. 39

Acetobacter   Lactobacilli

Ropiness

When a wine exhibits this condition it can almost, or in very bad cases, completely gel, or have in it a thick cord-like growth. The bacteria which causes this is a small sphere like organism called leuconostoc, which attacks any cane sugar (sucrose) present to form acetic and lactic acids. Leuconostoc mesenteroides because of its production of lactic acid is generally grouped with the other lactic acid bacteria. It is anaerobic as are many of the other lactic acid producing bacteria.

Detection of bacterial growth or infection

If for any reason a wine is suspected of having a bacterial contamination proceed as follows as soon as possible. Add 1 gm. (as much as will cover a 5p piece)—of sodium metabisulphite *immediately* to every two gallons of wine.

Take two glass 30 cc. (30 mls.) bottles with screw tops or rubber stoppers, a small glass funnel and a (60 mls.) bottle without a stopper. Sterilise by either boiling in water for 15 minutes or heating in a pressure cooker for 15 minutes at 15 lbs. per sq. inch pressure. Allow to cool and then put the funnel into the 2 fl. oz. (60 mls.) bottle and fold a No. 1 Whatnam filter paper to fit it. Pour some boiling water through the paper to sterilise it and discard the washings from the bottle, allow to cool for a few minutes and then filter a sample of wine through the paper. Well mix by swirling the filtered sample and divide it between the two sterilised bottles. Place one of the bottles in boiling water for a period of 15 minutes with the top slightly loose, and upon removal tighten the top down. Stand both bottles in an airing cupboard or other warm place and examine after 3, 5 and 8 days. If after this time, the unheated bottle is cloudy and the heated one still clear, a bacterial infection exists and another dose of 1 gm. sodium metabisulphite per two gallons of wine should be added. If, however, the two bottles are both clear the instability is unlikely to be micro-biological.

Prevention of bacterial attack or infection

The bacteria which may be encountered in wines are all susceptible to sulphur dioxide and if an addition of 100 parts per million or 1 gm. of sodium metabisulphite per gallon of wine is made early in the life of the juice little danger from bacterial attack exists. Under these conditions the malolactic fermentation will not take place and, as said above, the value of this to amateur winemakers is more than offset by the risk.

If for some reason the addition of sulphur dioxide via bisulphite has been omitted and a bacterial infection is suspected the full dose of 1 gm. sodium metabisulphite per gallon of wine (or juice) should be made immediately and not delayed until the results of the diagnostic test above are known. If the test shows a bacterial infection was present then, on the eighth day, a low dose of bentonite can be added to fine the wine bright again. The chapter on fining should be consulted to establish the dose necessary by means of a simple fining trial.

Do's and Don'ts

DO add a dose of sodium metabisulphite to your juices as explained in earlier chapters.

DO at all times work cleanly; in this way risk of micro-organism contamination is reduced to a minimum.

DO NOT despair if you have no microscope—a few simple tests will help you just as much.

DO NOT take panic action by throwing it all away if a mould colony is seen growing on a juice; skimming off the mould and fermenting quickly with a fresh starter may well prove successful in saving the juice.

DO NOT be tempted to omit dosing with sodium metabisulphite; it definitely helps to keep the fermentation clean.

DO NOT wait until the results of a test are known if a bacterial contamination is suspected, add some sodium metabisulphite at once.

CHAPTER 14

Troubles and cures

IN other chapters many of the troubles which can beset the home winemaker have been detailed and ways of preventing them explained. This chapter deals with the situation from a different angle and assumes a defect exists, the cause of which may or may not be known but has to be rectified.

Deciding the cause of the trouble

In many cases experience will tell the winemaker the nature of the problem, but for those lacking the necessary experience, a systematic approach can be recommended.

Basically all problems arising in home-made wines can be classified into two groups.

1. Micro-biological—in which case a bacterial, yeast or mould growth is responsible for the difficulty,

   or

2. Chemical—where the problems are ones of lack of balance, over sufficiency or a change in the chemical state of one or more of the constituents.

The first thing to decide is to which of these categories a problem belongs.

Classifying troubles

The following tables are intended to enable quick identification of a problem from its symptoms and, if possible, carry out a corrective treatment. The methods of test described in chapter 16 provide, in many cases, the basis for the identification and, the techniques of fining explained in chapter 7, the means of correction where practicable.

TABLE 6

Defect	Likely Cause	Treatment
A—Those due to micro-organisms		
Oily film on the surface of the wine.	A bacterial infection of probably acetobacter (vinegar bacteria).	If severe, apart from re-fermenting the wine, little can be done, however, if no alteration to the taste has occurred add 1 gm. (enough to cover a shilling) of sodium metabisulphite to each gallon of wine. Make sure the vessels are topped right up with wine.
Silky sheen when the wine is swirled.	This is likely to be due to lactic acid bacteria.	Well mix in 1 gm. of sodium metabisulphite to each gallon of wine and set aside for 2–3 weeks. A fining with bentonite might then be advantageous to prevent further instability.

Table 6 cont.

Defect	Likely Cause	Treatment
Viscosity of wine has increased or a thick rope-like growth has appeared.	This is most likely to be due to another type of lactic acid bacteria.	Using three lengths of clean cane bound together at one end, whip the wine vigorously to break up the growth and lower the viscosity, then immediately add 1 gm. of sodium metabisulphite per gallon, stand for 3–4 weeks in a sealed vessel and finally fine with bentonite.
A haze accompanied by small bubbles is seen in the wine.	Secondary yeast fermentation or Malolactic bacterial fermentation.	Either allow the fermentation to complete by removing the stopper and replacing it with a fermentation trap or add 1 gm. of Sorbistat (potassium sorbate) to each gallon of wine.
Small white patches on the surface of the wine.	This is probably due to an aerobic yeast growth sometimes called flor. Some moulds however can have similar appearance.	Carefully insert a racking tube so as not to disturb the growth. Rack the wine into a clean sterile container leaving all the growth behind, the sacrifice of a little wine is much more desirable than having some of the growth transferred. Add 1 gm. of Sorbistat to each gallon of wine as a further safeguard. On no account add sodium metabisulphite, as this will provide sulphur dioxide to combine with the aldehydes formed by the yeast, and will delay considerably the process of maturation. Make sure all vessels are topped up.

Table 6 cont.

Defect	Likely Cause	Treatment

B—Of chemical origin

Defect	Likely Cause	Treatment
Very acid taste.	Possibly the acidity was not properly adjusted in the juice before fermentation.	Measure the total acidity as described in the Test Methods of chapter 16. Check the recommended acidity in chapter 3 for the particular style of wine. If the difference in acidity between actual and recommended does not exceed 0.1 % as citric acid the cautious addition of a little precipitated chalk from the end of a penknife may be tried. When the bubbling stops the wine should be well mixed and the acidity retested. The number of further chalk additions can then be decided. Unfortunately this treatment usually leaves the wine a little flat and sometimes tasting faintly of chalk.
A very astringent taste. (A drying of the mouth, especially the roof and around the teeth)	High tannin content.	**1. Red wines** Set aside in a cool dark place to age. If this has already been done fine with gelatine after preparing a fining trial tasting the results in addition to observing them. (See chapter 7 for details of the gelatine fining and trial techniques). **2. White wines** Proceed with trial and gelatine fining as outlined above.

195

Table 6 cont.

Defect	Likely Casue	Treatment
White wine tends to darken quickly	The wine has a high content of ortho- polyphenyloxidase enzyme and low sulphur dioxide anti-oxidation protection.	Perform the test in chapter 16—"A test to see if a white wine is adequately protected from oxidation," and if a positive indication is obtained add sodium metabisulphite as recommended.
White haze in White Wine.	1. Starch (see test for starch in chapter 16. Methods of test).	If positive add Amylozyme .100 at the rate of 5 gms. per gallon.
	2. Protein (perform the test for white wine protein stability–chapter 16).	If positive carry out a trial bentonite fining and fine accordingly. (See chapter 7).
	3. Pectin – (see test for Pectin, chapter 16).	If test is positive use a depectinising enzyme in the proportions recommended on the preparation available.
	4. Precipitate due to oxidation. Carry out "A test to see if a White Wine is adequately protected from oxidation," (chapter 16)	Should this test prove positive add sodium metabisulphite as recommended.
	5. Iron contamination (see chapter 16 "A test for Iron in White Wines").	Proceed with treatment recommended, viz, citric acid treatment—unless already present — together with bentonite fining.

Table 6 cont.

Defect	Likely Cause	Treatment
Haze in Red Wine.		It is difficult to separate these. Filter a sample of wine into two small bottles, stopper them and put one in the ice section of a domestic refrigerator and keep the other at room temperature. If after 24 hours the refrigerated sample is very cloudy compared with the room temperature one, set up gelatine and bentonite trial finings as explained in chapter 7.
	1. Protein/tannin	
	2. Tannin/colouring matter	
	3. Pectin (carry out pectin test in chapter 16).	If positive add pectic enzyme in the proportions recommended by the manufacturer.
	4. Iron or other metal	Difficult for home-winemaker to detect. Set up trial fining with gelatine and bentonite (see chapter 7) as this is often effective.
Metallic taint to the taste.	Iron or copper contamination. Usually a haze or deposit.	When the taste is affected in the case of either red or white wine the best course is to fine with either casein or gelatine having first prepared a fining trial to decide if the treatment has any effect. See chapter 7 for details of the trial fining materials.

197

Table 6 cont.

Defect	Likely Cause	Treatment
Crystals in the wine.	If tartaric acid has been used for acidity correction these are likely to be potassium hydrogen tartrate, otherwise known as cream of tartar.	Place the wine in a refrigerator as near to the freezer as possible for about two weeks then rack into a clean vessel.
Smell of bad eggs.	Due to hydrogen sulphide production by yeast during fermentation.	Rack the wine through the air (*not* in the normal way excluding air) and make sure the vessels are full.
The wine is too deep in colour. (Either Red or White Wine).	1. White Wine. Could be due to oxidation or the type of fruit used.	Add ½ gm. of sodium metabisulphite per gallon of wine and set up a trial fining with skimmed milk (see chapter 7).
	2. Red Wine. Most likely to be caused by rich coloured fruit in too prolonged fermentation contact.	Set up a trial fining with gelatine (see chapter 7).
Coloured haze.	1. White Wine. Perform the test for iron in a White Wine (chapter 16) Could be due to iron or copper.	If positive proceed as recommended in chapter 16 or if this is ineffective set up a gelatine trial fining and decide the best course from this.
	2. Red Wine. Difficult to be definite, could be iron, copper or tannin/colouring.	See if a gelatine fining will eliminate the problem (see chapter 7).

CHAPTER 15

Simplified wine chemistry

IT has already been said that all wines consist of many substances in addition to ethyl alcohol and some of these have been mentioned. This chapter describes in a general way the substances which go together to make wine.

The first thing to establish is the difference between organic and inorganic chemical compounds.

Organic compounds are those which have the element carbon as the "backbone" or basis to their structure with the element hydrogen giving close support. Oxygen and nitrogen are two other elements which occur quite frequently giving specific characteristics depending on how they are involved.

Inorganic compounds are those consisting of basic elements other than carbon although carbon may be present as a minority constituent.

For those without any scientific training the situation will perhaps be more clear if organic substances are thought of as those produced by living matter (plants, animals or micro-organisms) and inorganic ones as having mineral origins. From this it is easy to see why many of the substances encountered in wine are organic.

Organic constituents

Each carbon atom can combine with four other single groups, in ethyl alcohol two carbon atoms are joined together then three hydrogen atoms complete the first carbon whilst two hydrogens and an oxygen combined with a hydrogen atom complete the structure.

Oxygen requires two single groups to complete its structure, if only one is filled by a hydrogen the other is free to combine with the carbon chain structure. If the principle of four bonds to each carbon is used it is possible to see more clearly how the various constituents are built up.

$$H - \underset{\underset{H}{|}}{\overset{\overset{H}{|}}{C}} - \underset{\underset{H}{|}}{\overset{\overset{H}{|}}{C}} - O - H$$

The composition of wine

Some of the specific chemical ingredients of wine vary, of course, depending on the fruit and the acid or acids used but many are common to any wine and the following will serve to show the diversity and number of the *principal* constituents.

Organic	Inorganic
Acids	Water
Alcohols	Mineral salts
Aldehydes	Sulphur compounds
Amino acids	
Esters	
Enzymes	
Glycerol	
Ketones	
Proteins	
Pectins	
Sugars	
Tannins	
Colouring matter	
Carbon dioxide	
Flavouring compounds	

It is now possible to examine each of these groups and see how they are formed or arise.

Alcohols

It is the —OH group which characterises alcohols. CH_2OH is methyl alcohol, CH_3CH_2OH ethyl alcohol, CH_3-CH_2CH_2OH propyl alcohol and so on. When the carbon chain of an alcohol is long it is said to be a Higher alcohol.

Ethyl alcohol is formed by the enzymic fermentation of the starting juice together with added sugar and is the principal product of the fermentation.

Methyl alcohol is formed in very small quantities probably as the result of pectin hydrolysis during fermentation.

Higher alcohols are produced in minute amounts by the yeast cells during the course of fermentation and taken as a group are referred to as Fusel oil.

Alcohols can be oxidised to aldehydes and it is the oxidation of ethyl alcohol either by excessive contact with air or bacterial action that produces acetaldehyde which in turn is oxidised to acetic acid.

Glycerol

$$CH_2OH$$
$$|$$
$$CH\text{-}OH$$
$$|$$
$$CH_2OH$$

Compounds with more than one alcohol group (—OH) in their structure are called *Glycols*. Glycerol is such a substance and having three alcohol groups is called a trihydric alcohol.

Glycerol is produced during the course of fermentation as an essential part of the process but the amount remaining in the wine varies with the temperature of fermentation, the species of yeast involved and the degree of sulphiting in the juice. This latter point has been explained in chapter 5.

Glycerol has a sweetish flavour and an oily appearance, it gives mellowness to the taste of a wine and is responsible for the "curtains" left on the glass after the wine has been swirled.

Aldehydes

These compounds have the characteristic group —CHO which can be better written as.

$$-C{\overset{\displaystyle O}{\underset{\displaystyle H}{\big\Vert}}}$$

Because the oxygen is joined twice to a single carbon atom the structure is weak at that point and this is why aldehydes are easily oxidised to acids or reduced to alcohols. When acetaldehyde is oxidised in wine a darkening occurs.

Acetaldehyde is an important intermediate compound in the production of ethyl alcohol by fermentation (see page 52). In this instance a reduction process—removal of oxygen or addition of hydrogen—is taking place.

$$CH_3C{\overset{O}{\underset{H}{}}} + H \longrightarrow CH_3C{\overset{OH}{\underset{H}{\underset{H}{}}}}$$

acetaldehyde   ethyl alcohol

or $CH_3CHO + H \longrightarrow CH_3CH_2OH$

It is likely other aldehydes exist in wine but in very minute quantities and little information exists about their identity.

When sulphur dioxide is added to a wine it combines first with water and then the sulphurous acid produced combines with acetaldehyde to produce a bisulphite compound. This prevents oxidation of the aldehyde and explains the importance of sulphur dioxide as an anti-oxidant.

1st step SO_2 + H_2O ⟶ H_2SO_3
             sulphur   water   sulphurous
             dioxide   acid

2nd step

$$CH_3\overset{O}{\underset{H}{C}} + H_2SO_3 \longrightarrow CH_3\overset{OH}{\underset{H}{C}}-HSO_3$$

acetaldehyde   bisulphite compound

Acids

Grapes contain a high concentration of tartaric acid; apples, pears and many other fruits contain more malic acid whilst citrus fruits are rich in citric acid.

The characteristic group of organic acids is —COOH and the formula of some of the common acids found in fruit wines or juices are given below. Acetic acid although undesirable in wine in anything other than trace quantities is included, because of its simplicity, to show how increasing the carbon chain length and arrangement gives different acids.

Acids	Formula	Remarks
*Acetic acid	$CH_3 . COOH$	Has one acid group, is the acid of vinegar.
Malic acid	CH_2—$COOH$ | $H.C.COOH$ | OH	This is a dibasic acid—It has two acid groups—and is changed by bacterial action to lactic acid in the malolactic fermentation.
*Lactic acid	$CH_3CH(OH).COOH$	Only one acid group. The basic acid of sour milk, is produced by the malolactic fermentation and as a by-product of alcoholic fermentation.
*Citric acid	CH_2COOH | $C(OH)COOH$ | CH_2COOH	A tribasic acid—3 acid groups — which can be broken down by certain bacteria to form acetic acid. Because of its ability to complex with iron it is very useful.

*Succinic acid	$\begin{array}{c} CH_2.COOH \\ | \\ CH_2.COOH \end{array}$	This dibasic acid is very resistant to bacterial attack and always occurs in alcoholic fermentation, the quantities varying between $\frac{3}{4}$–$2\frac{1}{4}\%$ of the alcohol produced.
Tartaric acid	$\begin{array}{c} HC(OH)COOH \\ | \\ HC(OH)COOH) \end{array}$	The acid of the grape. It is dibasic and forms salts which can precipitate from the finished wine after it has been bottled.

The acids marked with an asterisk (*) are produced during the process of fermentation in addition to possibly being present in the juice.

All of these acids—with the exception of acetic acid—are termed *Fixed Acids* because they have little or no smell and are not removed by passing steam through the wine.

Acids are important because micro-organisms do not flourish in sufficiently acid media, a fact which was mentioned in chapter 12, and they therefore give the wine protection. With some acids this protection is only partial, the acids themselves being attacked, this is so in the case of citric and malic acids.

When a titration is carried out for *Total Acidity* it is the concentration of free acid which is measured but acids have the ability to combine with metals and form organic acid salts. If tartaric acid combines with potassium two possibilities exist:

(a) Both of the acid groups will combine with potassium to form di potassium tartrate

$$\begin{array}{c} H—C(OH)—COOK \\ | \\ H—C(OH)—COOK \end{array}$$

   or

(b) Only one of the acid groups combine and a "half salt," potassium hydrogen tartarte is formed

$$\begin{array}{c} H—C(OH)—COOK \\ | \\ H—C(OH)—COOH \end{array}$$

The important thing to observe is that it is the hydrogen (H) on the end of the acid group (COOH) which is replaced by the metal because it is this hydrogen which gives titratable acidity. When the hydrogen is replaced by a metal the acid character is lost and often the solubility changed; a soluble acid might be converted into an insoluble salt. This happens in the case of tartaric acid. Tartaric acid is soluble, di potassium tartrate is insoluble and potassium hydrogen tartrate relatively insoluble, therefore when tartaric acid is present during winemaking with the potassium always present in fruits from the sap fed to them by the plant whilst they were growing, some di potassium tartrate will be formed and precipitate out of the wine thus reducing its titratable acidity, because what was tartaric acid is being removed as the salt. Potassium hydrogen tartrate is also formed and whilst most of this will be precipitated due to its low solubility some will remain. If the temperature falls when the wine is later being stored the solubility will decrease still further and a deposit of potassium hydrogen tartrate crystals will be seen in the storage vessel or bottle.

Volatile acids

These acids can be removed from wines by passing a current of steam through the wine and advantage is taken of this in the laboratory examination of wine.

The principal volatile acid of wine is acetic acid. It is formed by the oxidation of alcohol and acetaldehyde, trace quantities always exist but in larger amounts spoilage is either in process or has already taken place.

Another volatile acid is propionic (CH_3CH_2COOH). it is produced only in spoiled wine, probably as a result of micro-organism attack. Butyric acid ($CH_3CH_2CH_2COOH$) occurs under similar circumstances.

Formic acid (HCOOH) is produced in trace quantities during fermentation and has been reported by Vogt to be higher in raisin wines.

Carbon dioxide (CO_2)

Although a gas produced during the course of alcoholic fermentation it is of importance because it dissolves in the wine and forms carbonic acid with the water present.

$$H_2O + CO_2 \longrightarrow H_2CO_3$$

Water   carbon dioxide   carbonic acid (exists only in solution)

Neither carbon dioxide or carbonic acid have any smell or taste but they produce a definite sensation on the tongue. Carbon dioxide in dissolving prevents oxygen being dissolved and therefore reduces the possibilities of oxidation.

Esters

As a wine ages alcohols and acids combine together to form esters. One unit, called a molecule, of acid unites with one molecule of alcohol, one molecule of water is lost and a molecule of ester formed. The ester produced of course differs with the alcohol and acid reacting.

$$CH_3COO\boxed{H \quad HO}CH_2CH_3 \longrightarrow CH_3COOCH_2CH_3 + H_2O$$

acetic acid   ethyl alcohol   ethyl acetate   water
(written backwards)

Actually the aroma of ethyl acetate is an indication of wine spoilage because acetic acid rapidly forms the ester.

Once esters were thought to be responsible for the bouquet of a wine but modern research has shown most of the esters produced in wine to have little or no aroma.

Succinic and lactic acids produce esters with relative ease whereas citric acid hardly esterifies at all.

$$CH_2COO\boxed{H + HO}CH_2CH_3 \qquad CH_2COOH_2CH_3$$
$$| \qquad\qquad\qquad\qquad\qquad \longrightarrow \qquad |$$
$$CH_2COO\boxed{H \qquad HO}CH_2CH_3 \qquad CH_2COOH_2CH_3$$

succinic acid   ethyl alcohol   ethyl succinate
              (2 molecules)

Ketones

These substances are only formed in minute quantities and they have the group $>C = O$ as their characterising feature. Micro-organisms which contain the enzyme Lipase

206

will convert fixed acids to the methyl ketone with one less carbon in its structure than the commencing acid. For example if butyric acid was reacting the ketone acetone would be produced.

$$CH_3CH_2CH_2COOH \xrightarrow[\text{(O)}]{\text{lipase}} CH_3COCH_3 + H_2O + CO_2$$

butyric acid   acetone water carbon
(4 carbons)   (3 carbons)   dioxide

The probable mechanism is:

$$CH_3CH_2CH_2COOH \xrightarrow{\text{(O)}} CH_3CH(OH)CH_2COOH \longrightarrow$$

$$CH_3COCH_2COOH \xrightarrow{\text{(O)}} CH_3COCH_3 + H_2O + CO_2$$

Amino acids

These are now thought to be important in the development of wine bouquet and flavour although little information is as yet available. Amino acids contain nitrogen as an amine group ($-NH_2$) in addition to the now familiar $-COOH$ acid group. In nature they are of extreme importance as they are the basic units of proteins, a single protein molecule frequently containing twenty or more different amino acids. Fruits and fruit juices contain amino acids, the amount and particular acid predominating varying with the fruit involved and the conditions of cultivation.

The principle amino acids in juices or fruit wines are given below, many others have been identified and are likely to be just as important as those listed but these will serve as examples.

Name   *Formula*

Alanine

$$\underset{\displaystyle CH_2-CH-COOH}{\overset{\displaystyle NH_2}{\displaystyle |}}$$

Comments: About the same amount in juice and wine.

207

| Arginine | NH_2 NH_2 |

Arginine

$$NH_2 \qquad\qquad\qquad NH_2$$
$$|\qquad\qquad\qquad\qquad |$$
$$C-NH-CH_2-CH_2-CH_2-CH-COOH$$
$$\|$$
$$NH$$

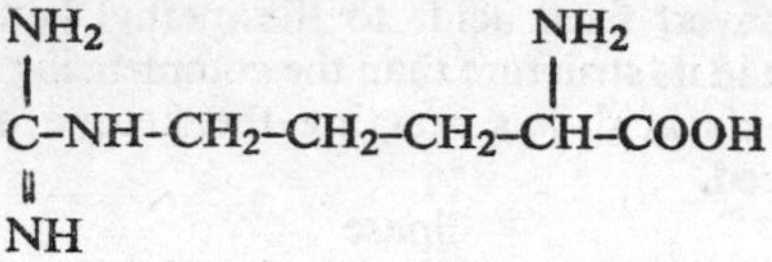

Comments: Less present in wine than in juice.

Glutamic acid

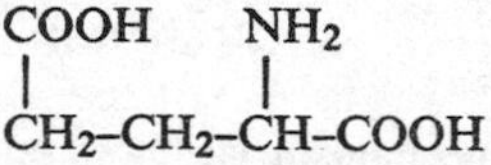

$$COOH \qquad NH_2$$
$$|\qquad\qquad |$$
$$CH_2-CH_2-CH-COOH$$

Comments: Present in both wine and juice.

Proline

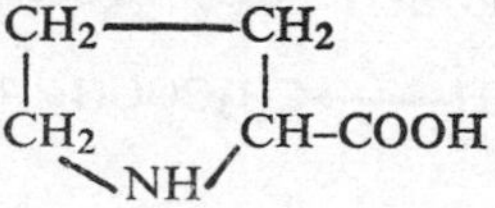

$$CH_2\text{------}CH_2$$
$$|\qquad\qquad |$$
$$CH_2 \qquad CH-COOH$$
$$\diagdown NH \diagup$$

Comments: Present in both wine and juice.

Proteins

It has already been said that amino acids constitute a large proportion of any protein, in fact putting the matter in very simple terms amino acids are joined together so that water is lost in the process and a —CO—NH— link, called a polypeptide linkage, is created. The amino acid alanine is used to illustrate this point below:

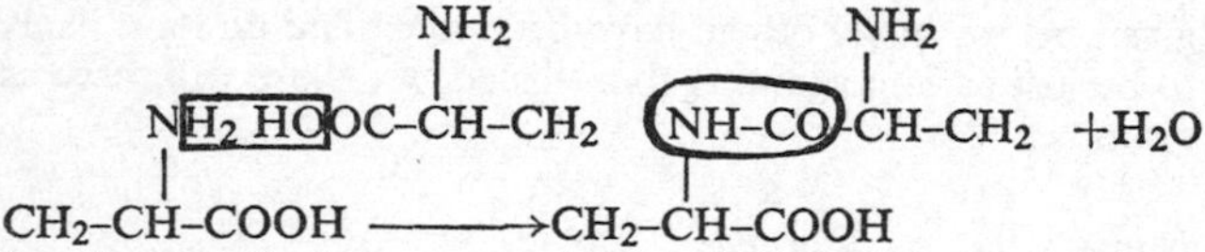

$$NH_2 \qquad\qquad\qquad NH_2$$
$$|\qquad\qquad\qquad\qquad |$$
$$N\boxed{H_2 \ HO}OC-CH-CH_2 \quad \boxed{NH-CO}CH-CH_2 \ +H_2O$$
$$CH_2-CH-COOH \longrightarrow CH_2-CH-COOH$$

2 molecules alanine   joined via *polypeptide link*

In a protein molecule many hundreds of these polypeptide linkages occur to form chains and the chains will form further bonds between each other creating a very complex chemical structure.

Some proteins are sensitive to heat or alcohol concentration whilst others are less soluble in water than in acid or alkaline solution.

Egg White (or albumen) when heated is denatured to form the familiar white substance everyone knows. This is an example of heat senstitivity.

When a wine has stood for a period of time a haze may form which cannot be attributed to any other cause than protein instability. In this case several factors are concerned including the change from an aqueous to an alcoholic solution, the highly reductive conditions during the process of fermentation and the oxidation changes subsequent to fermentation and storage.

Some proteins are also sensitive to cold and for this reason a wine exhibiting a haze may sometimes be rendered clear by placing it in a refrigerator for several days and then racking. What happens is the minute haze particles under the conditions of cold come together to form larger particles which are too heavy to be supported in the liquid and drop to the bottom of the container. Sometimes on warming up the reverse happens and the haze returns, this phenomena has not as yet been fully explained but is more likely to be due to tannin—cellulose complexes rather than proteins.

Certain proteins have physiological activity and are responsible for the process or initiation of biochemical reactions essential to life. In this category come nucleic acids which are present in the nuclei or governing centres of cells and are concerned with inheritance characteristics, and enzymes which promote all sorts of biochemical change.

Enzymes

The properties of enzymes concerned with fermentation are explained on pages 52-53.

Most enzymes are named either from the type of reaction they carry out or the compound upon which they act with the suffix "ase" added. Thus Oxidase carries out an oxidation reaction, whilst Maltase acts upon the sugar maltose. The yeast cells contain the enzymes which carry out the fermentation processes and thus as the yeast multiplies so the concentration of enzyme increases.

Pectolase acts upon the pectic substances and changes them to galacturonic acid or its methyl ester, as a result of this the partially alcohol soluble pectins are rendered soluble and do not give deposits in the finished wine. This enzyme reaction always takes place during fermentation but some fruits produce juices so rich in pectin that additional enzyme needs to be added so that a risk of later precipitation is completely eliminated.

Sucrase breaks down sucrose disaccharide molecules into monosaccharide molecules of glucose and laevulose which can then further react with other enzymes in the fermentation sequence as already explained on pages 51-52.

Amylase converts starch to maltose and dextrin which again is converted to glucose by maltase.

Carbohydrates

In this group are the sugars, starches, gums and pectins. The name is derived from the fact that these compounds contain only carbon, oxygen and hydrogen in their chemical structure, the first letters from the three words are used, viz. *CARB O HYDR* and *-ATE* added. Table 7 shows the structure of the more important members which are briefly discussed under the heading sugars.

Sugars

Glucose and laevulose

Simple sugars consist of single molecules and are called monosaccharides, these are then classified depending on the number of carbon atoms in the molecule. This will be more obvious if reference is made to Table 7. It will be seen that glucose is a monosaccharide hexose sugar; that is to say it is made up of single molecules each with six carbon atoms. Glucose and laevulose occur widely in vegetable and fruit matter the actual amount present varying with the individual species.

Sucrose

Cane sugar consists of molecules made up of two different monosacchride units joined together and for this reason it is called a disaccharide. The enzyme sucrase converts sucrose to glucose and laevulose although the process is often done artificially by acid hydrolysis—a boiling of the sugar with acid—in industrial processes, the resultant mixture is called "Invert Sugar" and is naturally present in honey. Under the acid conditions and in presence of yeast in any fruit or grape wine added cane sugar will hydrolyse without heating and be available for fermentation.

Very little sugar other than glucose (dextrose) or laevulose (fructose) exists in sound finished wine. Mannitol—a sugar glycol—is formed in wines attacked by mannitic bacteria due to the decomposition of laevulose and is evidenced by a "bitterish" flavour accompanying any residual sweetness.

Starch

Chemically this is a more complicated structure made up of glucose units linked together with a series of methyl groups ($-CH_3$) joined at fixed positions. Depending on the starch source further cross linking between molecules of the basic structure can take place to a greater or lesser extent.

In plants starch is changed to sugar by various enzyme processes but yeasts will not perform this task. However acid hydrolysis will take place especially if an acid is mixed with the starch and the solution heated. Potato and other vegetable starches can be converted in this manner. An alternative method, and one to use where maximum retention of any volatile matter is necessary, is to add a mixture of maltase and amylase to the water extract. First the sugar maltase is produced and then this is converted to glucose.

Starch turns blue with dilute iodine solution and this reaction is used to detect its presence in precipitates or hazes.

Cellulose is a polysaccharide with a structure very similar to that of starch.

Pectins

These are polysaccharides which are found widely in fruits and also some root vegetables and are responsible for the setting or gelling of jams.

TABLE 7—Some carbohydrates, their structure and reactions

Name	Type and class	Structure
Glucose (or Dextrose)	Monosaccharide Hexose sugar	
Sucrose (Cane sugar)	Disaccharide Sugar (2 sugar units joined together)	glucose unit laevulose unit
Laevulose (or Fructose)	Monosaccharide Hexose sugar (The structure is different in the free form to that when combined as in sucrose).	

More simply	Reacting enzyme(s) or chemical process	Occurrence	Compounds formed
CHO │ CH.OH │ CH.OH │ CH.OH │ CH.OH │ CH_2OH	Fermentation see page 51	Sweet fruits Honey	C_2H_5OH ethyl alcohol CO_2 etc. carbon dioxide, etc.
(structural formula of sucrose)	Sucrase (sometimes called invertase)	Sugar cane Beetroot Fruits	Glucose and Laevulose mixture (both have formula $C_6H_{12}O_6$)
CH_2OH │ CO │ CH.OH │ CH.OH │ CH.OH │ CH_2OH	Fermentation	Sweet fruits Honey	C_2H_5OH ethyl alcohol CO_2 etc. carbon dioxide etc.

TABLE 7—_Continued_

Name	Type and class	Structure
Maltose	Disaccharide sugar (Two glucose units joined together).	glucose unit glucose unit
Starch	Polysaccharide Double (Glucose units joined in a specific ratio. The recurring unit is called amylose).	
Cellulose	Polysaccharide (Glucose units joined together but in a different ratio and manner to starch).	
Pectin	Polysaccharide (Complex build up)	Very complex and numerous forms; Contains α galacturonic acid & methyl α galacturonate

More simply	Reacting enzyme(s) or chemical process	Occurrence	Compounds formed
CH——$CH(OH)$— $\|$ $\|$ $CH.OH$ o $CH.OH$ $\|$ $\|$ o $CH.OH$ $\|$ $CH.OH$ $\|$ $\|$ $\|$ $\|$ $CH.OH$ o —CH $\|$ $\|$ $\|$ $\|$ CH——$C\,H$—— $\|$ $\|$ CH_2OH CH_2OH	Maltase	Breakdown product of starch	$C_6H_{12}O_6$ glucose
$(C_6H_{10}O_5)_n$	Amylase or acid hydrolysis	Widely in the plant kingdom Potatoes, Cereals, etc.	$C_{12}H_{22}O_{11}$ + maltose $C_6H_{10}O_5$ dextrin
$(C_6H_{10}O_5)_n$	Hydrolysis with concentrated acid followed by boiling	Widely in the plant kingdom.	
$\begin{array}{cc} CHO & CHO \\ \| & \| \\ H.C.OH & H\text{--}C\text{--}OH \\ \| & \| \\ HOC.H & HO.C\text{--}H \\ \| & \& \quad \| \\ HOC.H & HO.C\text{--}H \\ \| & \| \\ H\text{--}C\text{--}OH & H\text{--}C\text{--}OH \\ \| & \| \\ COOH & COCH_3 \end{array}$	Pectolase	Fruits Roots of plants	$C_6H_{10}O_7$ & galacturonic acid $C_7H_{12}O_6$ methyl galacturonate

In structure they are chemically complex and can only be briefly described as units of ∝ galacturonic acid together with its methyl ester, the different forms of pectin depending upon the source to which they owe their origin.

Yeasts contain an enzyme, pectolase, which converts the complex pectin structures into the basic galacturonic acid and methyl galacturonate units, but if a large amount of pectin is present the change may not be complete hence the good sense in adding a pectic enzyme to juice before it is fermented. The breakdown of pectins causes a decrease in the viscosity of a juice or wine and assists in the general process of clearing.

Gums

During the process of fermentation some gums are always formed: they are carbohydrates of very complicated structure due to a great deal of cross linking between the various units.

Generally gums do not constitute a problem for the maker of fruit wines but some dried fruits, especially apricots and peaches, do tend to give on rehydration fruit which is rich in gums and from which it is not easy to make clear wine.

Tannins

These substances are present in the seeds, skins, stems and flowers of plants. The relatively simple tannins such as cathechol, gallic and ellagic acids are responsible for the astringent flavour often found in red wines. Gelatine readily complexes with gallic and ellagic acids and for this reason wines which have been fined with gelatine are often observed to be less astringent and bitter in their finish.

Catechol

Gallic acid

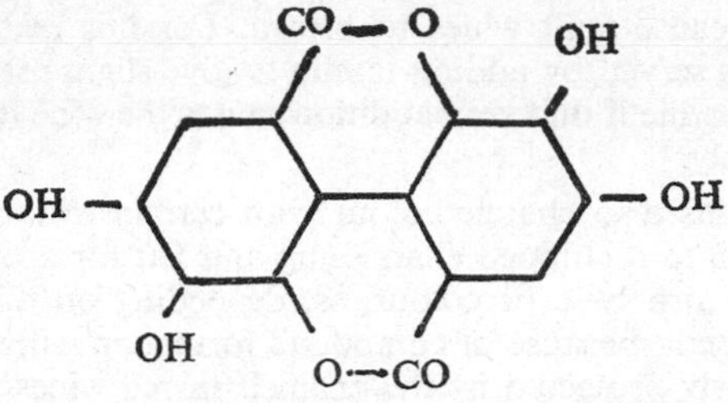

Ellagic acid

Tannins are often referred to as polyphenols, the reason for this is as follows. Phenol is a substance characterised by having one —OH group attached to a benzene ring structure. The benzene ring has a basic formula C_6H_6 but is more correctly written as:

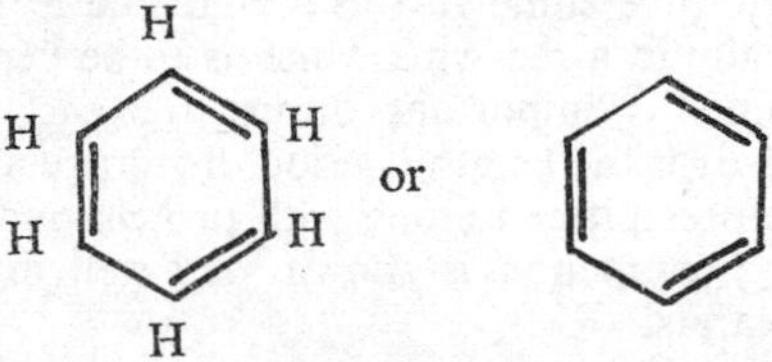

Benzene

Phenol is simply one of the hydrogen atoms replaced by the —OH or phenol group (only when the —OH group is attached to a benzene ring is it a phenol group, at other times it denotes an alcohol).

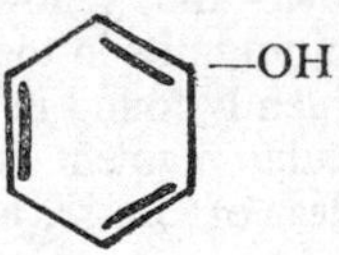

Phenol

When several phenol groups are attached to a benzene ring the compound is said to be a polyphenol.

Tannins having catechol as the basic unit of structure darken in the presence of a certain enzyme (polyphenoloxid-

217

ase) and cause fruit wines to brown. For this reason little purpose is served by adding tannin to give slight astringency in a white wine if that very addition causes the wine to darken prematurely.

Tannins also change colour with certain metals. A red juice put into a chipped enamel jug and left for a day or so may well turn blue or colourless, depending on the tannin forms present, because of compound formation with the iron not properly protected by the enamel. In red wines because, as will be seen later, tannins are closely allied to colouring matter, if excess iron is present it precipitates with the tannin and the wine will lighten in colour.

Similarly tannins form complex compounds with proteins and in doing so render the protein less soluble or denature it altogether resulting in a precipitate at the bottom of the wine.

Tannins do give some zest to a wine and a high tannin content especially in a red wine which is to be kept in bottle for a long time, is important during the early stages of maturation, as over the lengthy period of storage much of the tannin will be precipitated along with protein and colouring matter. This phenomenon is shown very well in the really high quality clarets.

Colouring matter

It has already been said that tannins are closely allied to plant colours and in fact they constitute a large proportion of the structure of many plant pigments although chlorophyll— the green pigment of plants, carotene—the red pigment of carrots and xanthophyll—the yellow colour of autumn leaves—do not have tannin units in their structure. Traces of these pigments are frequently found in juices and fruit wines, the amount and particular pigment depending on the juice source. Green vegetables, of course, are rich in chlorophyll, so are flower stems.

The other principal plant pigments are the anthocyanins which are red, blue or purple pigments and the anthoxanthins which are yellow, these all belong to a class of compound known as the *flavonoids* because chemically they have a close resemblance to flavone—a substance which occurs on

218

the stalks and leaves of the primula plant. The similarity can be seen from the formulae below:

Flavone

Chrysin
The pigment found in the buds of poplar

Delphinidin
Commonly found in fruit juices

Malvidin
Commonly found in fruit juices

Often these substances are combined with glucose sugar to give glycosides which are more soluble and abundant in nature. They are acid and alkali sensitive, the free antho-cyanins being violet in colour, the alkali salts blue whilst the acid salts are red, this accounts for the wide range of flower colours seen in a particular species of plant growing on different soils, a good example of which is the Hydrangea, and also for the fact that fruit wines contaminated with metal (commonly iron or copper) sometimes show a marked change of colour.

Some of the flavones exhibit the characteristics of tannins and are bitter to the taste, something which is noticed quite easily in Seville oranges due to the glycoside called naringin which does not occur in other oranges.

A further group of flavone compounds are the leucoan-thocyanins which are colourless but change to brown pig-ments. They occur in woody plant tissue, fruits and seeds. Tea, coffee and cocoa all contain leucoanthocyanins.

The chemistry of natural pigments is complex and what has been said in this book is but a peep into a vast subject all of its own.

Flavour components

Alcohols, esters, tannins, etc., all exert some effect to produce the overall flavour profile of a wine but other substances impart distinct flavour properties. Plants contain essential oils, which belong to a class of organic compounds known as the terpenes; lemon oil for instance which occurs in the peel of lemons is rich in the terpene citral. It also con-tains some limonene which is found to a larger extent in orange oil.

All of the essential oils have a pleasant odour or taste and since many are used as flavourings it is not strange, therefore, that they impart very definite flavours to fruit wines. It is reasonably true to say that any plant part which exhibits a pleasant smell contains terpene compounds (essential oils) and if the plant tissue is fermented along with the juice some of the flavours will be apparent either in the bouquet or palate of the finished wine.

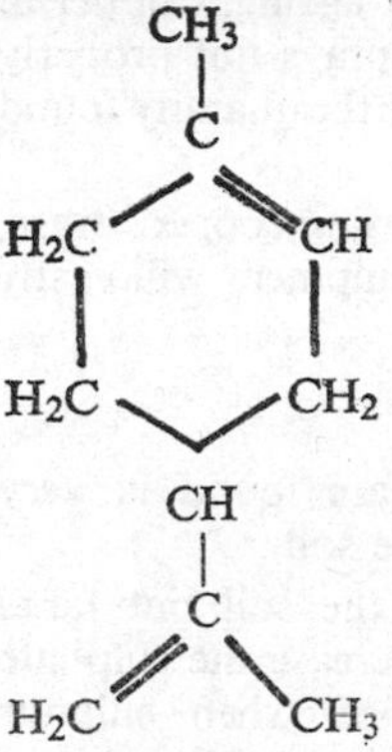

CH$_3$–C=CH–CH$_2$–CH$_2$.C=CH.CHO

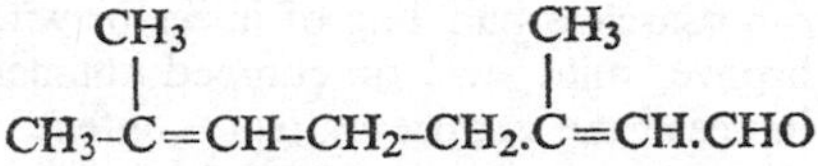

Citral (C$_{10}$H$_{16}$O)

Limonene (C$_{10}$H$_{16}$)

Inorganic or mineral constituents

Water

Something which is often overlooked is that on average between 85–90% of all wine is water. It should not be looked on with scorn, without it man would perish, but often the winemaker wishes he had a little less and proportionately more alcohol, not for reasons of bacchanalian joy but to show prowess in conducting a fermentation.

Metals

Many of these are essential in trace quantities for yeast growth. Sodium and potassium are always present in fruit wines, potassium contributing a greater percentage than all the other metals put together. Added sodium metabisulphite, of course, increases the sodium content of the wine but it does not affect the taste unless in excess.

Depending upon the soil from which the plant material came, magnesium, manganese, calcium, aluminium and iron are present to a greater or lesser extent, the quantity in any case being very small. During the process of fermentation the

221

amount of iron present in the juice is reduced considerably whilst the calcium content decreases as a result of salt precipitation either during fermentation or ageing. Copper in minute traces is often present but crop sprays not properly washed from fruit can considerably increase the quantity found in the finished wine.

Careless handling of juices or wines using copper, brass, bronze, mild steel or chipped enamel equipment will vastly increase the copper or iron content.

Salts

Chlorides, silicates and carbonates are found in very small amounts and owe their origin to the soil.

Sulphates also owe their origin to the soil but large amounts of added sulphur dioxide can increase the sulphate content because sulphurous acid, formed when sulphur dioxide dissolves in water, may be oxidised to sulphuric acid which has the sulphate group in its structure.

$$SO_2 + H_2O \longrightarrow H_2SO_3$$
sulphur dioxide   water   sulphurous acid

$$H_2SO_3 + \underline{O} \xrightarrow{\text{oxidation}} H_2 \boxed{SO_4} \leftarrow \text{sulphate group}$$
sulphuric acid

Phosphates are always present in juices and wines and for the reasons explained on page 51 they are essential to the process of fermentation. Fermentations carried out with fruit "pulp" present are enriched in phosphate material. Phosphates do not exist wholly in the free state some being combined with organic compounds, for instance with sugar to form glycerophosphates.

Sulphur compounds

Sulphur occurs in many proteins, which is fortunate as yeast relies upon some of the element being present in the fermenting juice. After fermentation however little sulphur remains from this source, most of what is present coming from the added sulphite.

When sodium metabisulphite is added to a wine or juice with the acid conditions present it first changes to sodium bisulphite and then liberates sulphur dioxide, the process can be seen from the equations below.

$$Na_2S_2O_5 + H_2O \longrightarrow NaHSO_3$$

sodium	water	sodium
metabisulphite		bisulphite
		(in solution).

and then:

$$NaHSO_3 + H^+ \longrightarrow Na^+ + H_2O + SO_2$$

hydrogen	sodium		sulphur
ions	ion		dioxide
(from the			
acids present)			

Some acidity is necessary to produce the hydrogen ions but this condition is always met in fruit juices and the wines made from them. The sodium ions produced combine with the acid which contributed the hydrogen ions to form a salt (see Fixed Acids). The sulphur dioxide produced then does one of two things,

1. It combines with water present and forms sulphurous acid, in which case it is said to be in the "free" state because sulphurous acid exists only in solution and the sulphur dioxide is easily available.

    or

2. It combines with sugars, aldehydes or ketones to form bisulphite compounds (see Aldehydes) in which case it is said to be "fixed."

Free sulphur dioxide gives antiseptic properties whilst that which is fixed functions mainly as an antioxidant preventing browning.

Due to the reducing properties of yeast during the process of fermentation sulphur dioxide is sometimes reduced to hydrogen sulphide (H_2S) gas which has an objectionable aroma of rotten eggs. Fortunately this is usually only **slight** and racking through air removes it, however in severe **cases** it is sometimes necessary to resort to carbon fining.

The meaning of pH

In the section headed Fixed Acids it was shown that the hydrogen (—H) at the end of the organic acid group (—COOH) gave the acidic properties and when it was exchanged for a metal ion a salt was formed and the acid character lost. As an approximation this is reasonable but is not quite what happens in fact.

When an organic acid or the salt of an acid is dissolved in water it separates (or dissociates) into ions. Salts dissociate completely but organic acids only partially. For example acetic acid when dissolved in water partially dissociates into acetate and hydrogen ions:

$$CH_3COOH \rightleftharpoons CH_3COO' + H^+$$

acetic acid $\qquad$ acetate ion $\qquad$ hydrogen ion

Whereas sodium acetate completely dissociates into acetate and sodium ions:

$$CH_3COONa \longrightarrow CH_3COO' + Na^+$$

sodium acetate $\qquad$ acetate ion $\qquad$ sodium ion

In chemical studies it is customary to determine the weight in grammes of a particular ion in one litre of liquid and express this as the number of gramme ions per litre.

Water slightly dissociates into hydroxyl (OH') and hydrogen ions ($H^{+'}$) and in very pure water the amount of hydroxyl ions exactly equals the amount of hydrogen ions, one ten millionth part of a gramme of each ion being produced in every litre of water.

Since the amount of each ion is the same it is possible to write:

$$H^+ = OH' = \frac{1}{10,000,000} = 10^{-7}$$

At this state of complete balance the water is neutral (neither acid or alkaline).

If the concentration of the hydrogen ion is multiplied by that of the hydroxyl ion a value called the *Dissociation Constant* is obtained. As the name implies this remains the same for water (at the same temperature) no matter what is dissolved in it. This value is:

$$\frac{1}{10,000,000} \times \frac{1}{10,000,000} = \frac{1}{100,000,000,000,000}$$

or

$$10^{-7} \times 10^{-7} = 10^{-14}$$

If an acid is added to water the effect is to add hydrogen ions and under these circumstances to maintain the dissociation constant at 10^{-14} less than 10^{-7} hydroxyl ions are required, similarly if an alkali is added to water the effect is to add hydroxyl ions and therefore less hydrogen ions are required from the water to maintain the dissociation constant. The overall effect in either case is to cause the water to dissociate *less* than it would if pure, so that as the added ion concentration increases so the other ion due to the dissociation of water decreases. It is logical therefore to express acidity or alkalinity in terms of either the hydroxyl or the hydrogen ion concentration since if the concentration of one ion is known the other can be calculated as the product of multiplying the two ion concentration together will always be 10^{-14}; the hydrogen ion concentration was selected, this is shown as [H$^+$].

The very small numbers involved have already been shown and the inconvenience is fairly obvious. To overcome this difficulty Sorensen, a Swedish chemist, suggested that the logarithm of one over the hydrogen ion concentration be used and called pH.

$$pH = Log \frac{1}{[H^+]} \quad or \quad -Log\,[H^+]$$

The advantage of this system is whole numbers with a maximum of two decimal places are used.

On this system the neutral point where the hydrogen ions equal the hydroxyl ions instead of being 10^{-7} becomes pH 7.0.

As acid conditions mean an increase in the number of hydrogen ions the acid range is from pH 6.99 — pH 0, alkaline conditions mean less hydrogen ions are produced by the water and the alkaline range is from pH 7.01 — 14.0.

pH in brief

pH is a scale based on logarithms which runs from 0–14 and gives a means of indicating the acidity or alkalinity of a liquid. The neutral point is pH 7, below this value down to pH 0 acid conditions prevail whilst above this value and up to pH 14 an alkaline state exists.

The important thing to remember about pH is that it is a logarithmic scale and a change of one unit up or down is a change of 10 times the acid or alkali concentration.

	pH	Grammes hydrogen ion H^+ per litre.
	0	1.0
	1	0.1
	2	0.01
Acid	3	0.001
	4	0.0001
	5	0.00001
	6	0.000001
Exactly neutral	7	0.0000001
	8	0.00000001
	9	0.000000001
Alkaline	10	0.0000000001
	11	0.00000000001
	12	0.000000000001
	13	0.0000000000001
	14	0.00000000000001

Buffers

Going back to the example of sodium acetate dissolving in water (page 224) a little more of the story can be told.

When organic acid salts of strong alkalis such as sodium acetate are dissolved in water they produce not neutral but alkaline solutions and the pH is not 7 but more of the order of 8.3, this fact has to be taken into account when selecting an indicator for the estimation of total acidity, the method which is described in the next chapter but a more important aspect of salt hydrolysis from the winemaker's point of view is the "buffering" action it has on the finished wine.

The organic acid salts have already been shown to completely dissociate whilst the acids themselves do so only partially. Now suppose a liquid containing acetic acid and sodium acetate was being studied, the ionisation would be as follows:

Salt, complete dissociation.

sodium acetate ⟶ acetate ions + sodium ions

Acid, only partial dissociation, some remains undissociated and therefore does not give an acid reaction acidity being a result of the hydrogen ion.

acetic acid $\quad$ partially $\searrow$ acetate ions + hydrogen ions.
(undissociated) $\searrow$ dissociates to

If an acid is added to the liquid the hydrogen ions of the new acid combine with some of the acetate ions of the salt to form undissociated acetic acid thus cancelling out the acid addition and maintaining a constant pH.

Similarly if an alkali is added to the liquid the hydroxyl ions combine with some of hydrogen ions from the acid to form undissociated water and the acid dissociates more to restore the ion balance. Again the effect on the pH is cancelled; this is known as "buffering."

Of course sodium acetate and acetic acid are not the acid and salt which are found in wine but the acids such as tartaric, citric, malic, lactic and succinic form small amounts of salt with the minerals present from the soil and behave in the same "buffering" fashion and cancel out the small acidity changes which take place during and after fermentation and hold the pH relatively steady.

CHAPTER 16

Methods of testing for amateur winemakers

Determination of specific gravity

Purpose

This test tells how heavy the liquid under test is compared with water, sugar and other dissolved solids increasing the value, whilst conversion of sugar to alcohol decreases it. It is possible then to determine the specific gravity of a juice and from the result form an opinion about the amount of sugar present. Similarly if the specific gravity of the juice is known and then at a stage during fermentation it is determined again from the difference between the first and second results an approximation as to how much alcohol has been formed can be made.

Method

An instrument called a Hydrometer is used for the test and fig. 40 illustrates one pattern in common use.

Basically it is a narrow glass tube sealed at one end and extending into a weighted bulb at the other. In the narrow tube is a scale with the highest value at the bottom and the lowest at the top. Thus it floats high for liquids with high specific gravity and low for those lighter. An important thing to notice is the temperature of calibration, this is either marked on the stem or the bulb and is the temperature at which the instrument scale was marked off. For the determination of alcohol content using specific gravity measurements it is particularly important to perform the test at each stage at the same temperature, because hot liquids have less density than cold ones, consequently if the test is performed at varying temperatures a serious source of error will exist.

A hydrometer with a scale 1.000—1.200 is useful for making good the sugar content of juices and watching the progress of fermentation but since alcoholic solutions have specific gravities less than 1.000 a second hydrometer with a scale 0.950—1.000 is necessary to assess the amount of alcohol formed in very dry wines.

Fig. 40

Apparatus required
Hydrometer
Hydrometer Jar or Measuring Cylinder
Thermometer 0–120°F. Stirring Pattern.

Technique
Fill the cylinder with the wine or juice to be tested to about (5mls) from the top. Measure the temperature with the thermometer and adjust the temperature by running warm or cold water over the outside of the cylinder, stirring with the thermometer at the same time, until the liquid is at 20°C. (or the temperature on the hydrometer).

230

Put the hydrometer into the liquid and by putting a finger on the top push it gently to the bottom of the jar, release the finger pressure and allow the hydrometer to rise. Read the scale at the *bottom* of the meniscus moving the eye down to that level as in fig. 41. The hydrometer must float freely and not rest on the side of the jar, if it does the resulting surface tension will give an incorrect reading.

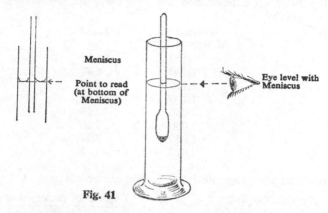

Meniscus

Point to read
(at bottom of
Meniscus)

Eye level with
Meniscus

Fig. 41

A determination of alcohol content by specific gravity difference

1. Where all of the sugar was added at the commencement of fermentation and fermentation was complete

Proceed as follows:

Determine the specific gravity of juice after the addition of sugar, taking care to adjust the temperature (see the determination of specific gravity) before carrying out the observation. Record the value. When fermentation is complete repeat the test.

The following calculation is based upon information given in a lecture by Dr. W. Honneyman in January, 1966, and is divided into three steps.

231

(a) First of all carry out the following calculation omitting the decimal points.

original specific — final specific gravity = degrees of
gravity of the   after complete   gravity lost.
juice   fermentation

(b) Look up a factor from the table below for the *starting* gravity

Original S.G. of sugared "must"	Factor
1.160	6.82
1.140	6.87
1.120	6.93
1.100	7.00
1.080	7.09
1.060	7.20
1.040	7.39
1.020	7.52

(c) The following calculation is then carried out:

$$\text{\% v/v alcohol in the completely fermented wine} = \frac{\text{Degrees of gravity lost}}{\text{Factor depending on starting gravity}}$$

Example:

If a juice had an original gravity of 1.080 after adding *all* the sugar, and when fermented fully the final specific gravity was found to be 1.002 the alcohol content of the wine may be calculated as follows:

Omit the deicmal points then

Original gravity — final gravity = degrees lost

$$1080 - 1002 = 78$$

Looking at the table a factor of 7.09 is found for a starting gravity of 1080

Thus

$$\text{\% by volume alcohol in the wine} = \frac{78}{7.09} = 11.1\%$$

2. Where the sugar has been added in stages during fermentation

This subject has been reported on by Mr. A. J. Jones, F.P.S., F.R.I.C., in a lecture at Liverpool in June, 1963.

Theory

Basically what has to be done is group the separate additions together and calculate what the starting specific gravity would have been had all the sugar been added at the commencement in one step. This is not so difficult as it sounds.

As already said in an earlier chapter when 1 lb. of sugar is added to 1 gallon of water or juice a volume expansion of approx. $6\frac{1}{4}\%$ takes place, which is 10 fl. ozs. Whenever a sugar addition is made the resulting expansion and the weight of sugar has to be taken into account.

The procedure is best explained by an example:

Initially 1 gallon of juice with a starting specific gravity of 1.065 was fermented and three sugar additions of $1\frac{1}{2}$ lbs., $\frac{1}{2}$ lb. and $\frac{1}{4}$ lb. made. How could the alcohol content after complete fermentation be calculated if the gravity was then 0.997?

Method of approach

1 gallon equals 160 fl. ozs.
Specific gravity $\times$ volume in fluid ozs. = Weight in ozs.

	Weight	Volume
1 gallon @ 1.065	$160 \times 1.065 = 170.4$ ozs.	160 fl. oz's
$1\frac{1}{2}$ lbs.	24 ozs. $1\frac{1}{2} \times 10 =$	15 fl. ozs.*
$\frac{1}{2}$ lb.	8 ozs. $\frac{1}{2} \times 10 =$	5 fl. ozs.*
$\frac{1}{4}$ lb.	4 ozs. $\frac{1}{4} \times 10 =$	2.5 fl. ozs.*
Total "original" weight	206.4 ozs.	volume 182.5 fl. ozs.

$$S.G. = \frac{\text{Weight in ounces}}{\text{Volume in fluid ounces}}$$

The "original" gravity would thus have been $\frac{206.4}{182.5} = 1.131$

*Since 1 lb. of sugar per gallon gives 10 fl. oz. expansion.

Comparison of proof strength with percentage alcohol by volume

Degrees proof spirit	% alcohol by volume	% alcohol by volume	Degrees proof spirit
10	5.8	2	3.5
11	6.3	4	7.0
12	6.9	6	10.4
13	7.5	8	13.9
14	8.1	10	17.4
15	8.6	12	20.9
16	9.2	14	24.4
17	9.8	16	27.9
18	10.3	18	31.3
19	10.9	20	34.9
20	11.5	22	38.3
21	12.1	24	41.8
22	12.7	26	45.3
23	13.2	28	48.8
24	13.8	30	52.3
25	14.4	32	55.8
26	14.9	34	59.3
27	15.5	36	62.8
28	16.1	38	66.4
29	16.7	40	69.9
30	17.2	42	73.4
31	17.8	44	76.9
32	18.4	46	80.4
33	19.0	48	83.9
34	19.5	50	87.4
35	20.1	52	91.1
36	20.7	54	94.6
37	21.3		
38	21.8		
39	22.4		
40	23.0		
41	23.5		
42	24.1		
43	24.7		
44	25.3		
45	25.8		
46	26.4		
47	27.0		
48	27.5		
49	28.1		
50	28.7		

All figures are correct to one place of decimals.

Omitting the decimal points,

Degrees of gravity lost = 1131 — 997 = 134

The "original" gravity has a factor from the table on page 232 of between 6.87 and 6.93 = 6.90

Therefore

% alcohol by volume in the finished wine =

$$\frac{134}{6.9} = 19.3\%$$

The determination of total acidity

Fig. 42

Many amateur winemakers are discouraged by the names of the chemicals and apparatus required for this simple test. This is quite unnecessary as the local chemist can easily obtain them and a laboratory supplier will have them ex stock.

A burette and pipette of the type recommended are shown below so that recognition will be easy.

Sodium hydroxide, or caustic soda as it is sometimes called, will remove paint if left on a painted or varnished surface and should therefore be washed off immediately if accidentally spilled. Similarly, if upset on clothes it should be washed off and a little vinegar used for the final rinse.

Phenolphthalein is used in medicine as a purgative and should not be allowed to come in contact with any food or drink for obvious reasons.

These comments are not meant to deter but to inform the reader. Simple precautions are all that are necessary and the information which can be obtained regarding the acidity of a wine or juice is very useful indeed.

Apparatus and chemicals:

> 1x25 mls. burette (a graduated tube with a stop tap, fig. 42).
>
> 1x5 mls. bulb type class B pipette (fig. 43).
>
> 1% Phenolphthalein indicator solution.
>
> N/10 sodium hydroxide solution.
>
> Distilled water (or condense water from a "fridge" when it is being defrosted).
>
> 1 glass beaker or tumbler. (250 mls. or ¼ pint capacity).

Method for white wine

Into the beaker put about 100 mls of distilled water (this is just under 4 fl. ozs.). Using the pipette measure 5 mls. of wine and run this into the water and add 5 drops of phenolphthalein indicator, swirl to mix.

Fig. 43

Hold the burette in a stand or by means of a pair of Terry clips screwed to a board. Fill the burette to the 25 mls. mark with N/10 sodium hydroxide, making sure the jet beneath the glass tap or clip is filled by allowing a little alkali to run out and then topping up.

Run the N/10 sodium hydroxide into the wine a few drops at a time keeping swirling to mix. When the first *pale* pink appears stop, wait a minute, and if the colour remains note the burette reading, if not add another drop. If the starting point was 25 mls. on the burette and after titration it is say 20.5 mls., then 25–20.5=4.5 mls. have been used, this volume is called the Titre.

Titre × 0.14 = Total acidity as % citric acid in the wine.

or

Titre × 1.4 = Total acidity as parts per thousand citric acid.

Rose or Red wines

The colour of wines interferes with the end point colour in the test above and it is necessary to spot drops of indicator out on a white tile to have an external indicator. In other respects the test is virtually the same. (Decolourising with charcoal tends to give results lacking precision).

Procedure

Take a 100 mls. quantity of distilled water in the beaker and add 5 mls. of the wine or juice just as before.

Using a *clean* white tile (a spotting tile—one with little depressions—is ideal, but a flat white tile will do) put at intervals of 2″ or so 2 drops of phenolphthalein. Have about twelve lots spotted out.

Fill the burette with N/10 sodium hydroxide and add about 10 drops at a time with mixing, then with a glass rod remove a couple of drops and mix with a spot of phenolphthalein. Keep adding and testing until a spot just turns pink. It may be necessary to rinse the tile and repeat the test adding only 2 drops of alkali at a time between the point where the first pink colour was noticed and the previous colourless spot to obtain an accurate result.

The calculation is the same as for white wines.

For those who work in part per thousand malic or tartaric acids:

Titre × 1.5 = total acidity as ppt. tartaric acid.

Titre × 1.34 = total acidity as ppt. malic acid.

Testing for pH

Purpose of test

To indicate the acid balance of the juice or wine being tested. Whilst the total acidity test tells how much "free" acid is present, pH also takes into account the acidity or alkalinity due to salts which may be present, this is explained under the heading of Buffers in chapter 15, and the result obtained is the balance of these factors.

Method of test

Liquid indicators are available but because of the colour range shown in home-made wines the results obtained are not always easy to interpret and the use of indicator papers is thus recommended. These papers are sold through all laboratory suppliers and a range from pH 2.8–3.8 is required. All that is necessary is for an indicator paper to be dipped into the wine or juice and the colour compared with that given for the respective pH values.

Interpretation of results

Even the most acidic wine should not be less than pH 2.9 or the acidity is likely to be objectionable, if this does happen a corrective treatment can be carried out as described in chapter 14.

If a wine is shown to have a pH above 3.5 acid should be added to bring the pH down, as above pH 3.5 bacteria grow more readily and spoilage is therefore more likely.

Determination of free sulphur dioxide

Purpose of test

To ensure that about 50 ppm. (or slightly more) free sulphur dioxide exists in a sound wine just prior to bottling. This will prevent any bacterial spoilage after the wine has been bottled.

Limitations

Because a colour change is involved the method is only suitable for white wines. Other methods exist for both red and white wines but involve more complicated apparatus and several chemical reagents.

Apparatus

1. A specially etched measuring cylinder marked in milligrammes per litre sulphur dioxide—this is the same as parts per million (fig. 44).
2. Bottle of blue solution (this is a special strength iodine solution with starch added together with a stabiliser).
3. A dropping pipette.

All the items can be obtained from H. Erben Ltd., Hadleigh, Essex.

Method

Fig. 44

Pour wine down the side wall of the cylinder until the liquid level just reaches line 0 with the cylinder held at eye level. Add a few drops of blue solution, close the cylinder with the palm of the hand and shake. Be careful not to allow any solution to spill as this will cause inaccuracy. Repeat the additions and shaking until a violet-red colour appears and stays for a few seconds after shaking. Read off the parts per million free sulphur dioxide direct from the cylinder. (Note. Blue solution remains stable for 8–10 weeks and it is therefore advisable to use it up amongst several winemakers rather than store it for a long period after which it is useless).

Action to take

If less than 30 ppm. "free" sulphur dioxide is present add to each gallon of wine just enough sodium metabisulphite to cover a sixpence and mix well.

Testing for pectin

Purpose of test

Pectic substances are always present in fruit juices but due to enzyme action during fermentation are in the main converted to other compounds which are either precipitated or rendered more soluble in alcoholic solution. Sometimes, though, not all the pectin is converted and because its solubility is less in alcohol than in water a haze forms in the finished wine. To prevent this happening a pectic enzyme can be added to the juice. If the treatment is suspected of being insufficient or forgotten altogether the following test may be used to show if pectin likely to cause trouble, is present.

Method

Using a large test tube, methylated spirit and a tea-spoon proceed as follows:

239

Place a teaspoonful of the suspect wine or juice into the tube followed by four teaspoonsful of methylated spirit and mix well.

A gelatinous haze which settles into a precipitate of the same nature indicates the presence of pectin.

Nearly always a slight opalescence occurs due to gums and resins which are present and this should not be confused with the indication of pectin.

Hazes

1. Testing for starch haze

Purpose of test

Sometimes when a fruit (or vegetable) containing a high starch content is used for winemaking some of the starch remains in the finished wine to cause a haze, with the aid of iodine it is easy to sort this type of haze from the many other types possible and deal with it.

Method

Take about 5 cc's. or one teaspoonful of the suspect wine in a test tube or small glass phial and add two drops of tincture of iodine, shake and examine the colour produced. A brown solution means no starch is present; a blue colour starch is present.

Action to take

If no starch is present test for unstable protein matter and pectin.

When starch is shown to be present the haze can be eliminated by adding amylase enzyme, the quantity required will vary with the preparation used and chapter 14 should be consulted for the approximate amounts.

2. Pectin haze test

See the method used under Testing for Pectin.

3. Testing for unstable colour or protein matter

Purpose of test

Red wines which are to be bottled often contain colouring matter which is rendered unstable and precipitated shortly after bottling. This type of instability is associated with oxidation during the bottling operation and can often be detected by a simple refrigeration test.

Method

Take two small screw top or corked bottles of about 15–30 mls. size.

Fill both bottles with the wine and stopper.

Place one bottle in the freezer section of a domestic refrigerator and retain the other at room temperature. After 24 hours examine both bottles after gently shaking. If the refrigerated sample is cloudy compared with that at room temperature the probability is that the wine does contain unstable colouring material.

Action to take

Either

1. Place gallon or half gallon jars of the wine in a domestic refrigerator for seven days and then rack immediately prior to bottling,
   or

2. Assuming large amounts of sodium metabisulphite have not been added previously add enough to cover a shilling to each gallon of wine followed by thorough mixing. This gives approximately 100 p.p.m of Sulphur dioxide.

4. Testing for a microbiological haze

See page 190 of chapter 12 for the method of test and the action to take on the results obtained.

A test to see if a white wine is adequately protected from oxidation

Often a wine darkens more quickly than desirable because the sulphur dioxide content is too low. Some wine-makers think an absence of sulphur dioxide is desirable. This is not so, sulphur dioxide prevents rapid oxidation by a mechanism described in chapter 15, page 202.

Test

Take about 100 mls. of wine and mix into it a dessert spoonful of kieselguhr diatomite filter aid. Next fold a filter paper as shown in chapter 7, and put it into a suitably sized filter funnel. Well stir the wine and kieselguhr mixture and pour it into the folded filter paper. Allow the wine to run through the paper into a glass and when it is all through discard it.

Now pour into the coated filter paper approximately 100 mls. of wine. Collect this wine in a suitably sized glass or bottle and either cover or cork it to prevent airborne contamination. Set aside for four days and then examine compared with a sample of the same wine taken from the storage vessel.

If the aerated sample is more yellow or brown to that from the storage container add 50 ppm. (or about as much as will cover a new penny) of sodium metabisulphite powder to each gallon of wine.

An easy test for white wine protein stability

Sometimes wines made with a high proportion of fruit can, if not fully matured, exhibit a fine haze due to protein instability soon after bottling. A simple test will tell if your wine is likely to suffer from this defect.

Test

Using a large test tube or small clear glass bottle take about 25 mls. of wine and add to it $1\frac{1}{2}$ mls. of 5% tannic acid solution (5 gms. of tannic acid dissolved in 100 mls. of distilled water). A medicine spoon graduated in ccs. or mls. is useful for this last addition or alternatively a 2 ml. graduated pipette can be used.

If a cloudy or flocculent deposit is formed it is likely that protein instability exists, a further test is then necessary to confirm this. If no haze or deposit is formed then the wine is protein stable.

Confirmatory test

To a 100 ml. sample of wine add 2 drops of 10% sodium metabisulphite solution (10 gms. of sodium metabisulphite dissolved in 100 mls. of distilled water) followed by 1 ml. of 5% Bentonite solution. Mix well, seal the bottle or tube and stand for 24 hours or until settlement has taken place then decant a clear sample and test with 5% tannic acid as before. If no cloud or deposit is now formed the original wine did have protein instability.

Treatment

To each gallon of wine quickly add 45 mls. of 5% Bentonite suspension and mix well. Allow to settle and then rack off.

A test for iron in white wines

If a metallic taint, white or blue haze is exhibited by a wine it is likely iron contamination exists. Other metals can be responsible but the detection of these is beyond the scope of the home winemaker. If iron contamination is only slight the following will help quickly to identify it and allow an attempt at treatment. Should the contamination be high, as is the case usually when the taste is affected, there is little the home winemaker can do.

(N.B. For satisfactory results the following tests should not be performed in direct sunlight).

Test

1. If a haze or deposit is present in the wine

Take a sample of wine in a clean test tube or bottle and add to it a pinch of sodium metabisulphite. Mix well and leave to stand for a few hours.

If the haze or deposit disappears but a deposit of yellowish colour appears on the bottom of the vessel, iron is very likely to be present.

2. **To establish remedial action necessary if iron is shown to be present or the taste is slightly metallic.**

Using a 25 ml. sample of wine add 1 ml. of 5% tannic acid solution, mix and see if a haze or deposit is formed.

(a) **No haze or deposit formed**

Perform the following tests.

(i) To 100 mls. of wine add two drops of 10% sodium metabisulphite solution and $\frac{1}{2}$ ml. of 5% citric acid solution.

(ii) To 100 mls. of wine add two drops of 10% sodium metabisulphite solution and 1 ml. of 5% citric acid solution.

When settlement has taken place filter each portion through a folded filter paper and set aside in covered glasses or corked bottles for four days. On the basis of which sample is clearest decide which treatment is best, remembering that the lowest acceptable addition should be used whenever possible.

If (i) is successful, add 45 mls. of 5% citric acid solution and 2 mls. of 10% sodium metabisulphite solution to each gallon of wine. Allow to settle and then rack the wine into a fresh clean container.

Should (ii) be successful, add 90 mls. of 5% citric acid and 2 mls. of 10% sodium metabisulphite solution to each gallon of wine.

(b) **A haze or deposit formed by the tannic acid solution**

Take 2 × 100 ml. samples as before and add to each 1 ml. of 5% Bentonite suspension followed by thorough mixing.

After half an hour use the two samples and proceed exactly as for (i) and (ii) above.

Whichever treatment is most effective to each gallon of wine add 45 mls. of 5% Bentonite suspension half an hour before the addition of the sodium metabisulphite and citric acid solutions in the amounts indicated.

Residual sugar test

Object of test

When a wine is nearing the end of its fermentation there is little activity and it is easy to think that fermentation has stopped. If wine is bottled with greater than 0.2% fermentable sugar a definite risk of secondary fermentation exists unless it is fortified or a special addition of potassium sorbate is made. Therefore a simple test to show the presence of more than 0.2% sugar is useful to guide the course of action necessary.

Apparatus required

2 Pyrex glass test tubes

1 bottle of Clinitest tablets for estimating sugar in urine.

(Price 20p. from any chemist).

1 Clinitest dropper.

Clinitest tablets are manufactured by Ames Company, Slough, who supply with each bottle of tablets a colour comparison chart for use in urine testing; it is this which is used in this test on wine.

Method

Make sure both test tubes are clean and dry.

Rinse out the dropper and into the first test tube place 15 drops of water.

Suck up some of the wine to be tested and place 10 drops into the second test tube. Wash out the dropper thoroughly and add 5 drops of water to the same tube. By gentle shaking mix the contents of the tube.

To both tubes add a Clinitest tablet. Wait for the bubbling to stop and then after 15 seconds shake the tubes gently.

The tube containing water only should give a blue negative (0%) colour on the comparison chart.

If the colour indication for the wine is that for $\frac{1}{2}$%+ or greater more than 0.2% sugar exists unfermented. The $\frac{1}{4}$% colour would indicate *less* than 0.2% sugar.

For storage of the tablets follow the directions at the bottom of the instructions given with each bottle of tablets.

This chapter has sought to put within the grasp of the home winemaker more scientific methods to determine the best course of action. Designing these tests to suit the home has of necessity imposed restrictions and for this reason these tests must not be seen as comprehensive. It is, however, the author's earnest hope that, even with the limitations, some use will be made of the methods described.

INDEX

INDEX

INDEX

INDEX

INDEX